Southern Space Studies

The Southern Space Studies series presents analyses of space trends, market evolutions, policies, strategies and regulations, as well as the related social, economic and political challenges of space-related activities in the Global South, with a particular focus on developing countries in Africa and Latin America. Obtaining inside information from emerging space-faring countries in these regions is pivotal to establish and strengthen efficient and beneficial cooperation mechanisms in the space arena, and to gain a deeper understanding of their rapidly evolving space activities. To this end, the series provides transdisciplinary information for a fruitful development of space activities in relevant countries and cooperation with established space-faring nations. It is, therefore, a reference compilation for space activities in these areas.

The volumes of the series are peer-reviewed.

Annette Froehlich
Editor

Outer Space and Popular Culture

Influences and Interrelations, Part 3

Springer

Editor
Annette Froehlich
University of Cape Town
Rondebosch, South Africa

ISSN 2523-3718 ISSN 2523-3726 (electronic)
Southern Space Studies
ISBN 978-3-031-25339-3 ISBN 978-3-031-25340-9 (eBook)
https://doi.org/10.1007/978-3-031-25340-9

This Springer imprint is published by the registered company Springer Nature Switzerland AG
The registered company address is: Gewerbestrasse 11, 6330 Cham, Switzerland

Contents

Mexican Mars: Narrating Spatial Futures from the Margins

Anne Warren Johnson

Abstract

This chapter looks at a series of events and encounters organized by the Mexican space art and science collective Marsarchive.org between 2020 and 2022. The workshops *Imagining Martenochtitlan*, *Martenochtitlan: Myth, Rite and Site* and *Mars in Guererro* were developed with the goal of creating speculative narratives and visual representations of imagined Mexican ("Neo-Tenochca") settlements on Mars. Focusing on the collaborative processes, national iconography, and contemporary preoccupations from which the workshops emerged, the text analyzes the ways in which these encounters and their products offer alternative visions of human futures in outer space, in contrast to the neocolonialist and extractivist proposals of representatives of space agencies and the private space sector.

1 Marsarchive.org

In 2016, a few years after the creation of the Mexican Space Agency, the International Astronautical Congress (IAC) was held in Guadalajara. Well-known astrocapitalist Elon Musk gave the keynote speech, "Making Humans a Multiplanetary Species," in which he invoked the planet Mars as "possible…something that we can do within our lifetimes."[1] He argued that humanity can follow one of

[1] "Making Humans a Multiplanetary Species." *YouTube*, uploaded by SpaceX, 27 September, 2017, www.youtube.com/watch?v=H7Uyfqi_TE8 (all websites cited in this publication were last accessed and verified on 26 July 2022).

A. W. Johnson (✉)
Universidad Iberoamericana, Mexico City, Mexico
e-mail: anne.johnson@ibero.mx

A. Froehlich (ed.), *Outer Space and Popular Culture*, Southern Space Studies,
https://doi.org/10.1007/978-3-031-25340-9_1

two paths: stay on Earth and, one day face extinction, or leave Earth and become a space civilization and multiplanetary species. The second path, Musk insisted, is the only way to survive. "I hope you all agree with me, yes?" The audience yelled and clapped. The image of Mars appeared behind him, a red sphere filling the screen. "This is what we want." More cries and applause.

Musk spoke of his desire to construct a self-sustaining city on Mars, referencing Mars' closeness to Earth and its geophysical characteristics, as well as the changes that Mars would have to undergo for humans to be able to find water, thicken its atmosphere, plant crops, and build a city. He explained how to achieve this dream. Right now, he said, a voyage to Mars would cost ten billion dollars per person. It would be difficult, but costs could be lowered through a series of technological innovations. He explained the technical requirements of rockets and ships, fuel composition and the infrastructure that would be necessary to maintain a Mars system. He presented a digital simulation scored by inspiring music, demonstrating the how the SpaceX Mars rocket launch and landing would work. The rocket's design would be based on the Falcon rockets and Merlin motors, and the ship would be based on the Dragon vehicles that already exist. At the end of the presentation, the digital planet revolved, transforming itself from a red, dessert sphere to a red, blue, and green sphere with an atmospheric halo indicating its inhabitability. This last image, a disquieting hybrid of Mars and Earth, accompanied Musk for the rest of his speech.

To sustain a civilization on Mars, Musk warned, we will need a population of one million people. With a hundred people per trip, the quantity of cargo, passengers and crew that will fit into the enormous rocket, it would take ten thousand trips to fulfill this dream, and, he estimated, it would take about one hundred years. Eventually, there will be "many ships…a thousand ships or more waiting in orbit, so the Mars Colonial Fleet would take off *en masse*. A little like *Battlestar Galactica*, if you've seen that thing. Good show." And Musk considered that the cost of a trip to Mars, including life support and the consumption of food, could end up being less than one hundred thousand dollars. "It's not just a dream, but something that can be made real."

Marcela Chao, museologist, and cultural promotor, saw the conference on YouTube, and was captivated by Musk's passion, if not his billionaire persona. In 2016, Chao was working in Mexico City at the Center of the Image on projects that combined science, technology, and art. She founded the organization that would later become Marsarchive.org[2] with her friends and friends of friends as a way of intervening artistically in the photographs generated by NASA's Mars rovers, but the project eventually expanded to include reflections on "the reactivated space race" led by space agencies and the private sector. Just in that moment, Elon Musk came to Mexico and, as Marcela recounts, "I got archive fever."[3] She decided

[2] See the organization's website, "Marsarchive.org: Vivir Marte desde la Tierra," www.marsarchive.org.

[3] Interview with Marcela Chao, Amadís Ross and Juan Claudio Toledo, 1 December 2020.

that Marsarchive.org would compile all existing information about Mars, like a contemporary version of the General Archive of the Indies in Seville, to "generate accountability" because everything is changing so fast in the space industry and, "how long before the millionaire gets bored, right?" She contacted various space actors in Mexico, included the Mexican Space Agency (AEM), the Center for Digital Culture (CDC), and the Space Generation Advisory Council (SGAC),[4] and Marsarchive.org began to grow and expand. An interdisciplinary group formed around Chao, made up of students and professionals in, among other fields, art, literature, engineering, and astronomy. Between 2017 and 2018, Marsarchive.org presented events such as the curatorial program MartePop, a Wikipedia editathon, a series of projections of movies about Mars and a virtual reality exhibition. They continued with the first edition of the Christmas *Posadas marcianas*, a series of podcasts and *La Bendita primavera marciana* ("Blessed Martian Spring") during the annal celebrations of Yuri's Night that commemorate Yuri Gagarin's voyage to space in 1961.

Some members of the group had more "scientific" perspectives, and so decided to "go down another road," while within the group, according to Marcela, "the craziness kept going like we wanted it to." The collective started to think about other topics, like the passage of time on Mars. To celebrate the Martian new year, which occurs every 687 days, they created the Martian Calendar, a collaborative project led by Marcela and Juan Claudio Toledo, astrophysicist at the National Autonomous University (UNAM). The dual calendar was meant to be a representation of the Martian "Year 35" which took place between March 2019 and February 2021 of the Gregorian calendar. It was illustrated by images of Mars obtained by NASA rovers, curated by Marcela, and explanations of the areological conditions of the planet written by Juan Claudio. The project was meant to reflect on how, on the one hand, time is determined by our origins in the solar system and how, on the other, time is a cultural construction influenced by political and historical realities.

Martenochtitlan was born after a discussion between Marcela and Juan Claudio about the possibility of founding a Mexican city on the site of the Mars landing of the rover Curiosity and was visualized as a "Neo-Tenochca futurist meme."[5] They decided that the meme deserved further attention and began to develop a workshop that would take up the possibilities of a Mexican inhabitation of Mars. Would the first Mexican Martian city be built on a lakebed racked by earthquakes, like the construction of Tenochtitlan and, later, Mexico City? And thinking about

[4] The Mexican Space Agency (AEM; aem.gob.mx) was founded in 2012 as part of the Secretary of Communication and Transportation; the Center for Digital Culture (CCD; centroculturadigital.mx) is part of a project sponsored by the Mexican government to promote digitally mediated artistic productions and disseminate technological and artistic knowledge. The Space Generation Advisory Council (SGAC; spacegernation.org) is an international organization for young people interested in participating in the space industry. They maintain an active group in Latin America.

[5] "Tenochca" is the word for the inhabitants of the Aztec capital Tenochtitlan, the site of present-day Mexico City.

Fig. 1 Codex "Martenochtitlan: Mito, Rito y Sitio", creative commons

Elon Musk's plans for Mars, should we terraform Mars or should we transform ourselves?[6]

2 Imagining Martenochtitlan

In 2020, Marcela and Juan Claudio met up with their friend Amadís Ross, a researcher at the National Institute of Fine Art and Literature (INBA) and expert in science fiction. Until then, Marcela told me, everything they did with respect to Mars was "universal," but "really when I say universal, well, everything was very American, you know?" For Marcela, this "universal" perspective would change through her participation in the seminar Aesthetics of Science Fiction, organized by the INBA's National Center for Plastic Arts Research, Documentation, and Information (CENIDAP), and coordinated by Amadís, who was convinced of the importance of a decolonial view, and "creating science fiction from where you are." According to Marcela, this perspective was fundamental because marsarchive.org was meant to imagine futures somewhere else, "although this future wasn't realist, I mean, one hundred percent scientific." Amadís was inspired by the name Marcela and Juan Claudio had given to their imaginary city because, "the name Tenochtitlan has great power in the Mexican imaginary…it touches the heart of Mexicans." The group is aware of the problems generated by focusing on the center of Mexico and the Aztec image to the exclusion of other pre-Hispanic populations: it could have been, they say, "Marte-Chichén Itzá." But, as Amadís recounted, other names wouldn't have the same mythical or sonic force. "Martenochtitlan" opens the door for an appropriation of the mythic symbolism of the nation's center, allowing for "dialoguing with the universe" and imagining new futures (Fig. 1).

They decided to present the workshop in three parts: in the first, coordinated by Amadís, participants would work with the mythic aspects of the narrative to be produced; in the second, they would develop the architecture and urban planning of the imagined city; and finally, they would write up a social contract to guide the Martenochcas in their new civilization. However, by the end of 2020, they had only been able to fully develop the first part of the workshop, which was sponsored by the UNAM's Program for Art, Science and Technology (ACT). Around sixty

[6] I was able to observe the second and third editions of the workshop. For the description of the first edition, I rely on conversations with the organizers and access to the myth and codex.

people responded to the call for participants, and twenty were selected, considering participants' diverse backgrounds in terms of discipline, gender, and age, although the majority were urban, educated, and middle class. Fifteen people ended up in the workshop, where they heard Juan Claudio talk about "the reality of Mars," emphasizing the planet's inhospitable nature, difficult terrain, dust, radiation and, above all, its lack of easily available water. "Everything you need to know if you're going to live on Mars," he said, so that the myth constructed by the participants would contain at least some reference to physical and technical realities.

Workshop participants also read the myth of the founding of Tenochtitlan recorded by Ángel María Garibay en his work *The Literature of the Aztecs*, as well as other literary texts, including *Invisible Cities* by Italo Calvino, Ray Bradbury´s *Martian Chronicles* and Kim Stanley Robinson's *Mars Trilogy*. They reviewed information about the planet's topography in order to plan out possible routes, obstacles and settlements. The goal was to write a mythic narrative and illustrate a codex that would offer an alternative vision of a human future on Mars, imagined, and produced from Mexico.

The first decision the group had to make was how to respond to the question posed by the organizers: Do we want to "colonize" Mars? The almost-unanimous answer was "absolutely not." Collective feeling went against what Juan Carlos characterized as "what is happening in space discourse all over the world, especially in the United States, becoming more neoliberal than ever." Workshop participants decided that the narrative of the Tenochca migration to Mars should show the Martenochcas "cohabiting, learning to live with the planet," without falling into colonialism or militarism.

> Before our times, when we only lived on the Old Island, little by little our ancestors were making Mother Earth, and themselves, sick. They had lost the divine *elhuayotl*, the soul that connects us with the world. Then the gods Caktikaktéotl, god of the void, Ehécatl, god of the wind, and Teohyotica Xalli, goddess of dust, came to us through portentous signs and charged us with a great mission: find a new land in a far-away place, where we could make communion with them and recover the *elhuayotl*. In the new land the chosen founded Martenochtitlan, a great city of reflection and purification. During this voluntary exile, the travelers would fight the demons of the world as eagle warriors, in order to rediscover their eternal essence. Through misery and sacrifice, distance, and nostalgia, they would cleanse their hearts and then return to the Old Island, carrying their message of peace and joy to its inhabitants. The new world would save the old one from itself. Mother Earth would be able to breathe again.[7]

The organizers divided the participants in three groups to work on each "act" of the myth. In the act called "The Voyage," the first group would imagine the immigrants' journey from Earth, the Old Island, and arrival on Mars, the Red Island. Their objective was to describe what would be the foundation of a new city on Mars from a particularly Mexican perspective and with a decolonizing vision.

[7] Marsarchive.org, "Martenochtitlan Primera Fundación", 2020, https://www.artecienciaytecnologias.mx/martenochtitlan.

Participants decided that, after hearing the call of their gods, "the great lords" would need to build a radically different ship than Elon Musk's Dragon, with its mythic European nomenclature and phallic structure. The future Martenochcas created "The Great Quetzalcóatl, the leaden plumed serpent, an enormous ship with golden metallic plumage that would cross the obsidian rivers that flow between the cosmic islands."[8]

The ship would make three voyages, the first two crewed by robots whose task was to find water and build the foundation of the imagined city; the third would be the beginning of the human settlement of Mars. After the first trip, the robots informed "their masters" that they had found frozen water in the Hellas Basin, and so the humans undertook their voyage. "The great lords were sure of the power of their science and their ingenuity. Their arrogance would follow them to the new land."[9] Four hundred passengers (the *Centzontili,* twenty groups of twenty, a sacred count for the Mesoamericans, and the basis of their numerical system) were chosen to undertake the voyage to the Red Island. Despite their careful planning, when they arrived, they realized that there was not sufficient water under Hellas' surface. Therefore, the *Centzontli* were forced to search for a new place to establish their city. They sent the machines to the four cardinal directions to determine the most propitious site.

Group number two worked on this phase of the narrative titled "The Pilgrimage." They wrote that, of the four machine-lead expeditions, "three were lost in the dust storms, and nothing was ever heard from them again."[10] The only hope lay to the north, so the *Centzontli* went to work building tunnels to shield themselves from radiation as they traversed the planet. "The bravest of them" traveled on the surface so they could scout ahead. To protect themselves, they used suits made of nopal and communicated with positronic devices transported by designated *tameme*, or carriers. Eventually, they narrate, the communicators acquired an "almost divine voice."[11]

During the voyage, the pilgrims encountered new life forms: small bacteria with which they underwent "a kind of fusion, or mestizaje."[12] This fusion caused a sickness that affected their mucus membranes and bones, transforming human bodies in "marzipan statues that ended up falling apart." The Martenochcas interpreted the disease, which resulted in the death of half of their population, as a punishment from the gods. Between this and other calamities, twenty-five years passed: "Some people died, others were born and grew up; these were much better adapted to the conditions of the Red Island."[13]

Eventually, the pilgrims arrived at Gale Crater, where they came across the ruins of SpaceX City, a settlement that had been founded by "the multimillionaire

[8] Ibid.
[9] Ibid.
[10] Ibid.
[11] Ibid.
[12] Ibid.
[13] Ibid.

Elon Musk and his followers." All its inhabitants had died of the disease caused by the Martian bacteria. "Perhaps the bacteria brought out what the inhabitants carried with them: colonialist ideas, war, resource expropriation, xenophobia, and misogyny."[14] Fortunately, they encountered in the ruins an anisble—an aparatus that made possible communication at speeds faster than light that was imaged by the U. S. writer Ursula K. LeGuin in 1966 —that let them know that there were natural resources available in the Plain of Tharis. (As Marcela said to me, "in the end, they couldn't completely let go of foreign science fiction".) The Martenochcas continued their pilgrimage, although many gave in to desperation. Only the consumption of a liquid that the bacteria produced, similar to pulque, the ancient, fermented drink made in Mexico from maguey plants, allowed them to survive. After another twenty-seven years, they came to the site upon which they meant to establish their city and burned the ansible in a ritual of purification. Unfortunately, the remains of the original *Centzontli* did not find resources on Tharsis and were forced to continue their journey. Fifty-two years had passed, a sacred Mesoamerican cycle.

The third group of workshop participants worked on "The Founding," the narrative's conclusion. They wrote that, inspired by a collective dream of Teohyotica Xalli, the goddess of dust, the pilgrims went to a cave, which they entered and thus began the last stage of their journey. The cave was dark, and the pilgrims were scared and uncertain. But a little girl spoke with the voice of the goddess and motivated them to continue. They found an ancient underground city, where they obtained fundamental knowledge of the planet's nature. After emerging from the cave, they crossed the rugged landscape of the Labyrinth of Noctis, and finally arrived at their ultimate destination: Valles Marineris, where they would build Martenochtitlan. "In honor of their origins, and as a means of united the place from which they had come and the place they had found, they erected a blue city, which contrasted with the red of Mars, a city in harmony with its planet. It was clear to the Martenochcas that they and the Red Island were one. There was no place for arrogance on this planet."[15]

In the last session of the workshop, the participants created the codex that would illustrate the myth they had created. Working from a collection of Mesoamerican glyphs that Amadís had provided them, participants chose the symbols that they deemed appropriate. In some cases, they created new icons, or fused together icons, to represent futurist notions such as the spaceship, but they maintained the aesthetic of their source material. Although as Amadís stated, they had the option to download Internet images, they decided not to "dirty" the original iconography.

---

[14] Ibid.

[15] Ibid.

3 Myth, Rite, and Site

The second edition of the workshop, *Martenochtitlan: Myth, Rite, and Site*, expanded on its predecessor's themes. Again, participants—mainly students and young professionals, whose backgrounds included art, architecture, literature, theater, psychology, graphic design, finance, computer programming, engineering, and chemistry—reflected upon pre-Hispanic myths and decolonial science fiction, learned about Mars' environment, and worked in groups to create a migration narrative and iconographic codex. And again, the question was posed, "What kind of world will we imagine?".

"We'll leave because we have to," suggested Quino. "Not just because we want to explore or colonize." Sandra agreed. "We should be like [the naturalist] Humbolt, and not Columbus."[16] The story began with an apocalyptic vision of Earth's future in which the planet had become desertified or "martianized" as the result the cataclysmic impact of a cosmic ray that was an ultimate consequence of human-caused climate change and excessive dependence on technology. Global communications broke down, and society collapsed, in a parallel with indigenous society after the Spanish conquest. But groups of indigenous and poor urban Mexicans, "Ehecatelpeños,[17] already "accustomed to crisis and scarcity"[18] organized themselves to take care of the few available natural resources. At the same time, there was a resurgence of traditional practice, and indigenous shamans once again became important sources of knowledge about the Earth and the cosmos. During nightly meetings, these leaders allowed people to once again "recognize themselves in the stars, remember that, before electricity lit up the night sky, there were stars, nature and spirit".[19] Rather than technology, cultural diversity and cultural *mestizaje*[20] became a source of strength and hope for the future.

Everything changed when a meteor fell to Earth, landing on the summit of a mountain in Mexico City. Eventually, it became clear that the stone was a message from the planet Mars, calling for an exodus from Earth, as the planet "needed a rest, needed a rest from us."[21] A seed came forth from the meteorite and grew into a giant nopal plant capable of space flight. And between mystical trance and advanced technology, the Ehecatelpeños boarded the nopal ship and traveled to

---

[16] Workshop *Martenochtitlan: Mito, Rito y Sitio*, August 2021.

[17] Ecatepec is an urban municipality adjacent to Mexico City that is considered one of the nation's poorest and most violent. Its name means "in the mountain of Ehécatl," the Aztec god of wind. The term "Ehecatelpeño" is a way of underlining the pre-Conquest roots of the place name and its inhabitants.

[18] Marsarchive.org, "Mito, Rito y Sitio", 2021. https://www.artecienciaytecnologias.mx/martenochtitlan-mito-rito-sitio.

[19] Ibid.

[20] The notion of *mestizaje*, biological and/or cultural intermixing between indigenous, African, and European populations, formed the ideological basis for Mexican identity from the beginning of the twentieth century.

[21] Ibid.

Fig. 2 "A Message from Mars". Detail from codex "Martenochtitlan: Mito, Rito y Sitio", creative commons

Mars, where they landed in the Labyrinth of Noctis and began a new era in their history (Fig. 2).

As was the case for the narrative constructed in the first taller, these Earth exiles also found a settlement, "with certain characteristics of a private corporation".[22] They encountered humanoid beings, Martian representatives of the corporation, who wanted to enslave them and subject them to genetic experiments. The Ehecatelpeños escaped, along with a "mestizo" Martian-Eathling hybrid, who betrayed them to the Martians. A series of battles, in which the Martians were guarded by the nocturnal gods Phobos and Deimos (versions of the two Martian moons) resulted in the death of many Ehecatelpeños. However, the mestizo came across a sacred cactus-like plant, ate its fruits, and was transformed into an avatar of Huitzilopochtli, Aztec god of war, becoming the protector of the Mexican immigrants. After a final bloody battle in which both Deimos and Huitzilopochtli perished, the Martians were defeated and returned to their city with the body of their fallen protector. Phobos decided to align herself with the Ehecatelpeños and was transformed into the Aztec goddess Coatlicue (Fig. 3).

The morning after they had entered the body of Huitzilopochtli, the Ehecatelpeños saw that the Martian land had become fertile, and from the god's grave spouted an ear of corn. They named the place Mount Huitzilopochtli, and there constructed their first settlement. Eventually, they flourished, and their lifespans were greatly extended. When they reached old age, the old men and women began to hear in the voice of Coatlicue the call of the landscape outside the city's walls, and they exiled themselves to the Martian desert, following trails of *cempasúchil* flowers,[23] the beginning of a spiritual quest and a practical way of maintaining

[22] Ibid.

[23] Cempasúchil is one way of writing the nahuatl word for marigold, a flower traditionally associated with the Mexican Days of the Dead.

Fig. 3 "Encountering the Martians". Detail from codex "Martenochtitlan: Mito, Rito y Sitio", creative commons

a fair distribution of their limited natural resources. After a period of hibernation wrapped in chrysalises, the elders would emerge as fully transformed Martians.

The narrative then describes a period of conflict between traditionalists who called themselves Rubrum and were tied both to Mars and their ancestral beliefs, and "technohumans" who believed in the power of metal and technology to lengthen their human lives, and thus avoid as long as possible their transformation into Martians. The Rubrum delved into the planet and constructed underground dwellings, while the technohumans were exposed to cosmic storms and affected by a fatal disease known as "the red fever." Eventually, the two sides negotiated a peace and exchanged technology and culture, reestablishing their city in the Plain of Argyre and adapting themselves "in a more ethical and efficient way" to their surroundings. They created new rituals to celebrate their union (Fig. 4).

As well as a myth and codex, participants in the second edition of the workshop created multimedia mockups of Martenochca architecture and descriptions of commemorative rituals accompanied by music playlists. They took up one of the most traditional representational forms of primary education in Mexico—the "monograph," a letter-sized infographic sheet about diverse topics, with a series of illustrations on one side and textual information on the other—to condense their ideas about the main Martenochca celebration that commemorates the major historical experiences, from their arrival in the nopal space ship to their conflicts with the native Martians, to their civil war and final reestablishment of Martenochtitlan (Fig. 5).

The impact of Martenochtitlan is not limited to the reproduction of well-known icons and literary tropes, but rather in their surprising temporal and geographic recontextualization. The workshop exhibited the indeterminacy of both the future and the past. In a Mexican key ("We aren't Mexicas anymore, but Mexic-anos," Amadís told me), workshop participants created (or appropriated) a series of "speculative technologies," to use the term proposed by Danish artist Kristina Anderson: Ursula K. Leguin's anisble, the spaceship Quetzalcóatl, the cactus suits, the energetic beverage similar to pulque, the artificial intelligence apparatus carried on the backs of the *tameme*, and others that I did not mention here, "embody a fear or a

Fig. 4 "Rubrum versus Technohumans". Detail from codex "Martenochtitlan: Mito, Rito y Sitio", creative commons

Fig. 5 "Reconciliation and celebration". Detail from codex "Martenochtitlan: Mito, Rito y Sitio", creative commons

desire in a form that allows me to try it out, live with its potentiality and rehearse living in this hypothetical future."[24] But technology is not the only thing humans will take with them to other worlds: as the third edition of the workshop shows, values and relationships may be even more important (Fig. 6).

[24] Kristina Anderson, "Making Speculative Technologies", in *Intersecting Art and Technology in Practice: Techne/Technique/Technology*, ed. Camille C. Baker and Kate Sicchio (New York and London: Routledge, 2017), 48.

Fig. 6 Martegrafía, "Martenochtitlan: Mito, Rito y Sitio", creative commons

4 Mars in Guerrero

The most recent edition of the workshop was held virtually, in conjunction with the Intercultural University of the State of Guerrero (UIEG).[25] The idea was to explore the future imaginaries of a more diverse group of participants, in this case, indigenous students (Nahua, Tlapaneco or Me'phaa, and Mixteco or Ñuu savi) from rural backgrounds. "It is easy," Amadís stated during the first session with students from Guerrero, "to imagine an astronaut in space eating a hotdog. It is harder to imagine one eating a plate of beans".[26]

Moving away from the geographical centrism implicit in the use of the term "Martenochtitlan" and considering the tensions (historical and contemporary) between the inhabitants of Guerrero and Mexico City, the encounter was called Mars in Guerrero, and participants were asked to pick a new name for their Martian settlement. The organizers repeated some of the themes of the first two workshops, including a discussion of the importance of myth and the presentation of information about Mars' physical characteristics, but after the first conversations in which

[25] Guerrero is considered one of the poorest states in Mexico for its low rates of income and access to health and education services, as well as high rates of violence.
[26] Workshop *Marte en Guerrero*, April-June 2022.

the students expressed their own interests, decided that the focus of this edition would be the creation of a social contract for life on Mars, a topic that, for reasons of time, had been left out of the first two editions.[27]

Interestingly, the indigenous students, mostly from the fields of sustainable development and forestry, although a few were studying language and culture, were much less interested in imagining a neo-pre-Colombian culture on Mars than their urban predecessors. The first names they proposed for the settlement did not reference indigenous culture at all; rather they spoke to more a desire for the establishment of a utopia in a new world: New Hope, New Mars, and Happiness. Other names did reference contemporary indigenous place names and notions: Martliaca (a combination of Mars with Atliaca, the hometown of several participants), Martli and *Biyu Natse* ("Eagle in the process of growing or being born" in Me'phaa). Aside from the inevitable references to Hollywood science fiction movies, much of the discussion revolved around the participants' main concerns: natural resources, social relations, and political organization. Students used their own experience to suggest the kinds of plants that might thrive in the absence of abundant water and to propose the kinds of habitats that might be sustainable on the surface of Mars or underground. They agreed that the social structure should be based on communalism, with rotating responsibilities and a series of rules and sanctions that would preclude the spread of corruption. Many were interested in the kinds of relations that could be established with any possible alien beings encountered, emphasizing the importance of mutual understanding, solidarity, and respect. They explicitly stated their desire to "not make the same choices as the Spanish conquerors," avoiding violence and domination.

Sadly, the workshop was cut short because of problems of time and communication: rural Guerrero continues to suffer from intermittent internet connectivity, and participants were unable to create the kinds of visual products that characterized the first two editions of the workshop. However, the elaboration of a set of principles for communal living on Mars is one of the more inspiring results of Martenochtitlan/Mars in Guerrero and represents a radical alternative to the current proposals circulating in the international space industry. Among these principles, participants state the following: live in harmony with other humans as well as Native Martians; establish agreements that emphasize respect and healthy human relations; provide mutual support and solidarity in the event of any crisis or accident; and establish rules that foment the value of environmental care and responsibility.

5 Conclusions

Martenochtitlan generated a space of imagination and narration and negotiated the production of series of visual and textual objects that established a speculative

[27] Ibid.

Mexican presence in outer space. This presence depended on the activation of a combination of what Diana Taylor has terms the "archive" and the "repertoire."[28] As iconographic historical narratives, Mesoamerican myths and codices are part of a Mexican culture archive, the inscribed register of a historical period, accessible in the present, separatable from its historical context and the agents of its production. But through its re-elaboration in the workshop *Imagining Martenochtitlan*, the founding myth of the Aztec city becomes an element of repertoire; it is implicated in a series of interactions and transformations that emerge from the interplay of new actors, knowledges, and interpretations. This transfer, as Taylor wrote in the context of the resignification of pre-Hispanic practice during the colonial period, amounts to a kind of performance that "allows for an alternative perspective on historical processes of transnational contact,"[29] substituting "interplanetary" for "transnational."

The civic education imparted to all Mexican students, whether indigenous or not, privileges a glorified pre-Hispanic past as part of national, mestizo (mixed indigenous and European), identity. Imagining Martenochtitlan restages the narrative and iconography of the Aztec pilgrimage that play a central role in this identity, largely thanks their representational condensation on the Mexican flag, with its central image of an eagle perched on a cactus with a serpent in its mouth, the sign that Huitzilopochtli gave to the Aztecs to mark the end of their years of wandering and show them where to build their capital city. But it also restages the dominant astrocapitalist discourse of the "conquest of space" represented by Elon Musk's vision of the human inhabitation of Mars and referenced in the narratives constructed by workshop participants in the form of the Martenochcas' encounter with the ruins of SpaceX City or the description of the Martian enslavers as Martian "corporations". The Mexican versions, based on the paradigm of "re-founding" rather than "conquering," participate in a subversive act of transfer; instead of a militaristic narrative of domination and extraction, Martenochtitlan and Mars in Guerrero substitute alternative ecological scenarios that emphasizes humility, auto-transformation, and interspecies collaboration. The narratives are messy, cobbled-together ideas that do not hide the organic, collective process through which they were constructed. They are neither linear, nor heroic; they leave no room for a messianic astrocapitalists to shape a shared future.

As Donna Haraway reminds us, "it matters what stories we tell to tell other stories with."[30] Imagining the human exploration of space as a story of conquest from the point of view of the historical conqueror has different consequences than imagining the human exploration of space as a story of cohabitation and negotiation from the point of view of those who have historically been conquered.

[28] Diana Taylor, *The Archive and the Repertoire: Performing Cultural Memory in the Americas* (Durham and London: Duke University Press, 2003).

[29] Ibid., 20.

[30] Donna Haraway, "Receiving Three Mochilas in Colombia: Carrier Bags for Staying with the Trouble Together", in *The Carrier Bag Theory of Fiction*, Ursula K. Le Guin (London: Ignota, 2019), 10.

Speculative fiction, while it may imagine futures, is also firmly anchored in the present. In Le Guin's words, "Science fiction properly conceived, like all serious fiction, however funny, is a way of trying to describe what is in fact going on, what people actually do and feel, how people relate to everything else in this vast sack, this belly of the universe, this womb of things to be and tomb of things that were, this unending story."[31] Martenochtitlan and Mars in Guerrero explore contemporary concerns: institutional corruption politicians' desire to preserve their power, immigration, the environmental crises that affect daily life at various scales, a sense of being left out of national and international decision-making. Imagining life on Mars, as Amadís Ross pointed out in one workshop session, means exploring the poetic (and political, I would add) implications of living on another world that will physically last longer than Earth, even when the sun goes supernova, "but still being tied to the same star."[32]

Telling a narrative of a human presence in outer space by means of particularly Mexican stories, Martenochtitlan and Mars in Guerrero reclaim a place for Mexico in the future and in outer space, but in a way that radically defies the stories told by astrocapitalists and interplanetary colonialists. In a ludic key, and not without conflict along the way, they imagine collectivity instead of individuality, self-transformation instead of terraforming. They remind us that, however "universal" it may seem, a vision of humanity in space is always a vision from somewhere.

Anne Warren Johnson is a Professor in the Graduate Program in Social Anthropology at the Universidad Iberoamericana in Mexico City. She received her MA and PhD in Social Anthropology from the University of Texas at Austin and holds a B.A. in Anthropology and Theater Arts from Brown University. Her research interests include the social study of science and technology, the anthropology of the future, performance studies, historical memory, and material culture, and she has published books, chapters, and journal articles in these fields. Her current project, based on ethnographic research with the Mexican Space Agency, a university space instrumentation laboratory, and a series of art collectives, revolves around Mexican imaginaries of outer space and the future.

[31] Ursula K. Le Guin, *The Carrier Bag Theory of Fiction* (London: Ignota, 2019), 37.
[32] Amadís Ross, workshop *Martenochtitlan: Mito, Rito y Sitio*, 10 August 2021.

Jellyfish from Outer Space: Tentacular Creatures and Cosmic Responsibility in Environmental Art and Pop Culture

Anna-Sophie Jürgens and Anne Hemkendreis

Abstract

Jellyfish from outer space appear in pop culture, scientific experiments and climate art, where they function as an expression of the tentacular and thus act as a key metaphor for ecological thinking and environmental responsibility. The tentacular is inextricably linked to the aesthetics of both the deep sea and outer space. It is a reflection on humanity's entanglement with nature and a symbol of pressing climate issues that permeate every aspect of our lives. The jellyfish from outer space concept and trope raises questions about the idea of leaving a planet ravaged by humans to its own devices, and about the planet as something beautiful to be protected. Precisely because jellyfish are, in a sense, formless creatures, they are transitory figures that can liquefy an established nature-human relationship and transform it into something new. Space-themed jellyfish are fluid figures of thought in art and pop culture that transfer moral questions of climate justice and climate awareness to new environments and solidify complex theories through their concrete reference to sensual forms of life.

Swimming while flying through a colourful galaxy, Gonzo meets two "cosmic knowledge fish", who know "many, many things" and call themselves "highly evolved beings". Although we unfortunately do not learn much more about their area of intellectual expertise, these educated cosmic fish from the 1999 comedy

A.-S. Jürgens (✉)
Dr. Anna Sophie Jürgens, Australian National Centre for the Public Awareness of Science, The Australian National University, Canberra, Australia
e-mail: anna-sophie.jurgens@anu.edu.au

A. Hemkendreis
Dr. Anne Hemkendreis, Research Group "Heroes, Heroizations, Heroisms", University of Freiburg, Breisgau, Germany

A. Froehlich (ed.), *Outer Space and Popular Culture*, Southern Space Studies,
https://doi.org/10.1007/978-3-031-25340-9_2

film *Muppets from Space* make a comeback 12 years later, albeit in a slightly different form. In the 2011 musical comedy *Happy Feet Two*, Bill and Will, two sophisticated krill, tell us a similar story: they compare the aesthetics of the submarine with that of the universe and discuss their own evolution and potential to move up the food chain. The passing of a luminous jellyfish is admired by the krill ("how beautiful") and becomes a symbol for the wondrous beauty of earthly life. Both films explore the aesthetic blending of two seemingly very different environmental spheres: deep space and deep sea.

Surprisingly, the aesthetic conflation of outer space and the underwater world is not that rare in popular culture. If you look closely enough, the melding of space and sea phenomena can be detected in the hippie reflection that we are all "particles in a cosmic jellyfish" from the 1971 TV film *In Search of America* (25:18), in James Cameron's 1989 science fiction film *The Abyss* (revolving around alien jellyfish), Ang Lee's 2012 adventure-drama *Life of Pi* (featuring a gleaming phosphorescent jellyfish), Disney's 2016 computer-animated musical *Moana* (presenting a glow-in-the-dark spaceship-like manta ray) and even Aqua Rick and Aqua Morty wearing fish heads and space helmets presumably filled with water (e.g. in the "Close Rick-Counters of the Rick Kind" episode of *Rick and Morty* (2013-)). More so, in James Cameron's *Avatar* franchise, flying and glowing jellyfish creatures play a prominent role. The film *Avatar 2—The Way of Water* (2022) shifts the well-known outer space action to an alien and volcanic underwater landscape; it exerts a great sensual appeal on viewers through the fantastic design of a weightless and iridescent world inhabited by mysterious, tentacled creatures. The aesthetic blending of underwater and space environments connects our world to a realm which is still mainly unknown, seemingly unpolluted and untouched by humankind. In times of climate change, this can be interpreted as a relational aesthetic and a sign of hope: the interchangeability of the deep sea and the cosmos ties our world—increasingly troubled by environmental catastrophes—to the vision of possible future environments, if not homes, for humankind.

Similar to (animated) film, art has worked with submarine/cosmic aesthetics to highlight humans' entanglement with the environments of the Earth, and beyond. The Center for Art and Media in Karlsruhe (Germany) showed the artwork *The Jellyfish* by Mélodie Mousset and Edo Fouilloux (2021) in the exhibition "Biomedia: The Era of living Media". Using VR headsets and microphones, visitors could communicate with luminous jellyfish—presented in the accompanying text as "non-human entities"—and thus experience "some kind of connection".[1] The visitors were immersed in a tranquil, dreamlike underwater world, populated by luminous jellyfish, and mesmerising bright stars and a large full moon. The tentacular creatures were attracted by the visitors' movements and voices (especially the imitation of whale songs): the aquatic creatures began echoing their sounds, communicating and joining in cross-species harmony in a seemingly outer space

[1] Zabludowicz Collection, "Introduction to 360: Mélodie Mousset and Edo Fouilloux", *Vimeo*, 2021, https://vimeo.com/639061568.

like environment. The effect was a feeling of dissolving one's own physical boundaries and a sense of emotional and material connection with an alien, yet strangely familiar, environment. These outer space jellyfish in (popular) art connect environmental theory with environmental sciences. They refer to a 1991 NASA science experiment (the first Life Sciences Mission) where 2000 moon jellyfish (this is, in fact, their real name) were sent to space in the space shuttle Columbia. Scientists found that baby jellyfish born in space later had problems adjusting to Earth's gravity—which means that human babies born in outer space might have the same issues. These experiments and reflections on the similarities and differences between seemingly zero gravity environments on earth and outer space environments are of increasing importance in times of climate change, as our planet is in real risk of becoming uninhabitable.

The ways in which submarine and cosmic aesthetics merge and inspire awe through otherworldly beauty and grandeur have not been extensively explored so far, although understanding these phenomena—for example, as metaphors for environmental reflection and messages—can deepen our comprehension of the intangible cultural aspects of environmental fragility and urgency.[2] Given our growing awareness that we need to protect our planet from anthropogenic climate change, it is time to ask how the manifold connections between cultural fantasies of outer space and the submarine can be defined and interpreted; where do human experiences of both environments intersect and how are they explored in popular culture and art? What do submarine creatures in space teach us about our experiences, cultural ideas and possible futures regarding earthly and cosmic worlds? What can we learn from jellyfish from outer space—as a leading metaphor of environmental thinking and cultural experiment—in terms of environmental awareness and a re-consideration of human-nature-relations? Through a close reading of examples from the field of pop cultural studies and arts, this chapter approaches the 'jellyfish from outer space' phenomenon through a series of adventurous theses at the intersections of aesthetic, historical and theoretical inquiry. The aim is to paint a suggestive—but by no means exhaustive—mosaic of ideas, themes and arguments around the influences and interconnections between outer space and deep sea in our cultural imagination.

1 Theoretical Framework and Relevance: Tentacular Creatures and Relational Aesthetics

Octopus or jellyfish-like creatures are found, almost inflationarily, in ecological theories as figures of thought about the relationship of humans to their environment. In other words, jellyfish appear in these contexts as a twofold figurative expression: they give visual expression to abstract theoretical concepts concerned

[2] Cf. Burns, T.W., O'Connor, D.J., and Stocklmayer, S.M. (2003). Science communication: A Contemporary Definition. *Public Understanding of Science*, 12, 183–202.

with our material entanglement with other living entities, and they function as symbols for the uncanny threat of climate change that is infiltrating every aspect of human life. Thus, jellyfish can be interpreted as both a sign of a deeper connection and an invisible threat. Donna Haraway talks about the primordial metaphor of the tentacular, which expresses situatedness and the connection of every being with another.[3] As a Gorgonian metaphor (a reference to Greek mythology), there is something unworldly and still anthropological about tentacular creatures. As a metaphor—of origins, a blueprint of possible worlds, their times, material-semiotic beings—the tentacular is an abstract cultural phenomenon *and* it has a concrete material counterpart in forms of nature, such as mushrooms, spiders and, indeed, octopuses and jellyfish.[4] Thinking with the tentacular rejects that there are independent organisms and systems. It also points to the fact that climate change has already left its mark on the deepest levels of the ocean, the highest layers of the atmosphere and in space.

Timothy Morten compares global warming to a tentacular horror creature that impacts all aspects of our lives.[5] At the same time, he calls for humanity to form alliances with the non-human—a radical intimacy that can change the relationship between humans and the world. He writes: "What explains ecological awareness is a sense of intimacy, not a sense of belonging to something bigger: a sense of being close, even too close, to other lifeforms, of having them under one's skin. (…) The proximity of an alien presence that is also our innermost essence is very much its structure of feeling".[6] Tentacular creatures serve as figures of thought for climate change; they embody human entanglement with the surrounding world and its life forms and establish a meta-reflection on the definition of the self in the encounter with "the Other" (the non- or more-than-human). In this sense, we argue, tentacular creatures are simultaneously material and immaterial. They inhabit the planet as (different) living beings and are also abstract reflections on the state of the planet and human connectedness and responsibility towards it.

In this regard, tentacular creatures—especially when they appear in popular culture and art—point to a sense of place as well as a transgression into outer space.[7] Ursula Heise predicts: "new possibilities of ecological awareness inhere in cultural forms that are increasingly detached from their anchorings in particular geographies".[8] The author traces our ecological awareness to the famous "Blue Marble" photograph from the 1968 Apollo 8 mission. Jellyfish-like beings in outer space

[3] Cf. Haraway, D. (2018), Unruhig bleiben. Die Verwandtschaft der Arte in Chthuluzän. Frankfurt, New York: Campus, 48.

[4] Ibid., 49.

[5] Morton, T. (2013), Hyperobjects. Philosophy and Ecology after the End of the World. Minneapolis, London: University of Minnesota Press, 71.

[6] Ibid., 139.

[7] Cf. Heise, U.K. (2008). Sense of Place and Sense of Planet. The Environmental Imagination of the Global. Oxford: University Press.

[8] Ibid., 13.

environments, however, communicate a very special kind of ecological awareness: instead of an imaginary standpoint in the cosmos which objectifies Earth, humans interact with submarine creatures in cosmic environments, thus projecting themselves into, and exploring, outer space. By showing submarine creatures in cosmic environments—detaching them from their environmental anchoring—films and art foster a trans-planetary thinking which makes the universe more familiar and potentially habitable for humans. In doing so, tentacular creatures become metaphors for the human need to overcome the limitations of their biological form.

This chapter draws on environmental philosophy and science communication, and refers to approaches from the emerging fields of Blue Humanities and Space Humanities. The Blue Humanities emphasises the idea of earthly networks in which each entity has its own rights and power to act. It investigates the interplay of human and non-human actors in cities and more natural areas. As Sidney Dobrin succinctly put it, the Blue Humanities are interested in the tension between post-human approaches and the question of global responsibility and cultural differences.[9] It is central to "acknowledge that any efforts to demonstrate entangled thinking risks obfuscating the role of human politics in all aspects of global ecologies. Blue ecocriticism negotiates a fluid space between and among the human and posthuman".[10] The Space Humanities is an even younger discipline, originated from the International Space University of Strasbourg (France) which launched its Space Studies Program in 2019.[11] It explores the significance of humankind's expansion into space with regard to ethical and ecological reflections. In doing so, it aims to understand the basic motivations for space activities by exploring their origins in, and their impact upon, human culture and society and by enabling new and renewed visions which can inspire future space programs. Referring to approaches from the Environmental, Blue and Space Humanities, our narrow focus on jellyfish in outer space examines how human culture already expands, and perhaps overcomes, its anthropocentric focus within the arts and popular culture (Fig. 1).

2 "Why Not Combine Outer Space and Inner Space?"—From Jellyfish on Spaceship Earth to Jellyfish Spaceships and Bioluminescent Cosmic Escapism

Electric jellyfish glow in the dark because they reflect the moonlight. At least this is how their eerie radiance is 'scientifically' explained in Wes Anderson's 2004 comedy-drama film *The Life Aquatic with Steve Zissou*, a parody of, and homage to, diving pioneer Jacques Cousteau. Remarkably, David Bowie's space songs accompany most of the protagonists' eccentric adventures on and under

[9] Cf. Dobrin, S.I. (2021). Blue Ecocriticism and the Oceanic Imperative. New York, London: Routledge.
[10] Ibid.
[11] Cf. Space Humanities (HUM), https://ssp18.isunet.edu/ssp18/9-space-humanities-hum.

Fig. 1 TA: *Technojellicus Cosmosis* 560 (2022). Orchestrated in MidJourney by TA

water,[12] where their submersible looks like a floating spaceship. The association between submarine and spaceship is no coincidence; it is often invoked in fictional and non-fictional films, and—perhaps surprisingly—jellyfish are never far away. "Deep ocean expeditions always seem like space missions" to deep-sea diver and film director James Cameron, for example, and are "way more exciting than any made-up Hollywood special effects". That is why Cameron teamed up with NASA scientists for his 2005 documentary *Aliens of the Deep* to investigate the "most insane alien life forms that have ever been discovered" on Earth, because "why not combine outer space and inner space?—why not take astrobiologists and space researchers [underwater]?"[13]

In 2012, when he explored the Mariana Trench with the purpose-built one-man submersible "Deepsea Challenger", Cameron returned to this comparison, relating his exploration of the deep sea to a spacecraft flight, the result of which "is the science and exploration and the imagery we're getting" (this was turned into the 2014 3D documentary *Deepsea Challenge*).[14] Delving into what has been called the "oceanic sublime"[15] of the last frontier for science and exploration on this

[12] Cf. Jürgens, A.-S. (2022). Being the Alien: The Space Pierrots and Circus Spaces of David Bowie, Klaus Nomi and Michael Jackson. *Southern Space Studies: Outer Space and Popular Culture 2*, Ed. A. Froehlich, Cham: Springer, 1–18.

[13] *Aliens of the Deep*, dir. David Cameron (DVD, all quotes 05:24–05:26).

[14] Boyle, A., "James Cameron Compares Deep Sea to Outer Space in Adventure Tale", *News*, 2012, https://www.nbcnews.com/science/cosmic-log/james-cameron-compares-deep-sea-outer-space-adventure-tale-flna742903.

[15] Hammond, B. (2013). The Shoreline in the Sea: Liminal Spaces in the Films of James Cameron. *Continuum (Mount Lawley, W.A.)*, *27*(5), 690, 690–703.

planet—the deep sea—Cameron was particularly fascinated by the lunar splendour he encountered, as well as the pulsating jellyfish.[16] Indeed, jellyfish seem to embody the kinship between the underwater world and extraterrestrial worlds, and not only in Cameron's non-fictional work. In Cameron's 1989 science fiction film *The Abyss*—whose visual sequences are partly reminiscent of *2001: A Space Odyssey*[17]—luminously tentacled jellyfish-like aliens resolve the threat to human civilisation posed by nuclear weapons; they represent intelligent life at the bottom of the ocean and Earth. These glowing and undulating purple-blue creatures have been recognised as a "prototype for *Avatar*'s colour scheme".[18] They are aesthetic precursors of the inhabitants of the lush habitable moon of a gas giant in the Alpha Centauri star system, called Pandora, which *Avatar*, Cameron's most successful franchise, is all about.

In fact, jellyfish fly through many galaxies of our fictional space narratives. In the *Star Wars* universe, they appear as Hydroid Medusas in the form of jellyfish cyborgs (enhanced by the Karkarodons with armour and cybernetics; prime material to be read in terms of Donna Haraway's environmental *Cyborg Manifest*). In *Star Trek* we come across a "Jellyfish Spaceship" from a future, foreign galaxy (a high-tech scout "Explorer Ship", designed as a science vessel). Not only that, but the *Star Trek* crew themselves come face to face with an alien spaceship that transforms into a dangerous jellyfish-like space creature (in the 1987 "Encounter at Farpoint" episode of *Star Trek: The Next Generation*). This seems to anticipate the tentacled, highly communicative extraterrestrials in Denis Villeneuve's 2016 science fiction film *Arrival*. In these pop cultural examples and Cameron's (non-fictional) documentaries, the transgressive figure of the tentacular jellyfish appears not only as a figure of thought about the relationship of humans to their habitat and their connection to other living beings (both on Earth and on an interplanetary scale); it also manifests as a nerve-centre in a pulsating network of identifications and confrontations of contrasting notions of inner and outer spaces.

We know about the fusion of deep and headspace—for example in the form of spiritually transcendent cosmic pensiveness—from the (con)fusion of adventures in, and imaginings of, outer space in the works of David Bowie.[19] More recently it emerged in Noah Hawley's 2019 drama *Lucy in the Sky*, in which outer space unsettles a NASA astronaut's inner space, visually epitomised by a scene in which

---

[16] Cf. Borenstein, S., "James Cameron, others to explore the real abyss", *PhysOrg*, 2012, https://phys.org/news/2012-03-james-cameron-explore-real-abyss.html.

[17] On Kubrick's influence on Cameron see: Howell, P., "James Cameron Had to Wait for Science to Catch Up to His Sci-Fi Ambitions." *The Toronto Star (Online)*, 2018, https://www.thestar.com/entertainment/movies/opinion/2018/04/26/james-cameron-had-to-wait-for-science-to-catch-up-to-his-sci-fi-ambitions.html.

[18] Tobias, S., "The Abyss at 30: Why James Cameron's Sci-Fi Epic is Really about Love ", *The Guardian*, 2019, https://www.theguardian.com/film/2019/aug/09/the-abyss-james-cameron-30th-anniversary-sci-fi-epic-love.

[19] Among other artists see: Jürgens, *Being the Alien*, 1–18.

the space traveller almost drowns in her spacesuit. There is, however, one contemporary novel—Sequoia Nagamatsu's 2022 *How high we go in the dark*—that puts a particularly interesting twist on this theme while also raising questions about our collective ecological consciousness. Nagamatsu's debut novel is not only a powerful reflection on the human experience of loss, despair and hope in the face of impending environmental collapse, but also an impressive tale of escapism in inner and outer spaces. His fictional world is partly inhabited by—you might have already guessed—jellyfish. Thus, at one point we read:

> All around us, spheres of iridescent light the size of hot-air balloons descend like a school of jellyfish. We are too mesmerized by the beauty of it to look away or even think about being afraid, as if we've been gifted with the sight of a cosmic occurrence like the birth of a star or the death of a planet, the aurora in a thousand snow globes.[20]

This metaphoric jellyfish instant occurs in a dreamlike reflection of a comatose protagonist whose planet (Earth) is being ravaged by environmental disasters caused by climate change. The scene is reminiscent of the liquescent deep space (dream?) sequence in Ang Lee's film *Life of Pi*, in which glow-in-the-dark jellyfish and a giant phosphorescent blue whale surround the eponymous hero in the inky waters of unfathomable oceanic vastness. Here, as there, the glamorous majesty of bioluminescent jellyfish—radiating light themselves—creates a living light interface and dreamscape around the protagonists, whether in the form of an underwater star system blurring into the night sky or a cosmos of outer space luminescence. In other words, cosmic jellyfish are mesmerising; both stories indicate that this admiration of jellyfish ultimately distracts from the catastrophes that surround them. Both stories are about human agency in the face of overwhelming forces that are incomprehensible and unchangeable for the individual, which in turn seems to be encapsulated by the jellyfish shimmering in the dark as a sign of both hope and escapism.

Escapism is the mental "diversion of the mind to purely imaginative activity or entertainment as an escape from reality or routine".[21] Escapism can be a coping skill, a psychic retreat, a form of distraction from feelings of sadness, depression or sorrow, but also the expression of the inability to engage and connect in meaningful ways with the world. In the interconnected, partly nested stories of *How high we go in the dark* (an intriguing structure reminiscent of David Mitchell's space-themed 2004 novel *Cloud Atlas*) space imaginaries and imagery, space costumes, pop cultural space references, stars and reports of nebulae and celestial phenomena, embody a dreamscape and an all-uniting, almost universal human longing to go beyond the devastated or devastating terrestrial, and human nature. Similarly, in Nagamatsu's novel, water-based 'jellyfish art' is another immersive escapist vehicle to 'leave' the planet imaginatively—and, surprisingly, even physically. In one

---

[20] Nagamatsu, S. (2022). How High we Go in the Dark. London: Bloomsbury, 70.

[21] "Escapism." *Merriam Webster Online*. https://www.merriamwebster.com/dictionary/escapism. Accessed 6 June 2022.

of the later chapters, a victim of a pandemic virus is turned into liquifying art: a melting human ice sculpture whose water tentacles run into the ocean. Thus, the jellyfish is a beautiful, if ephemeral, sign of inevitable loss and an evocation of the environmental awareness that everything in this world is interconnected.

In summary, within our filmic and literary fictions and visual pop cultural stories, the cosmic jellyfish—as a natural phenomenon and scientific object, figure of thought, narrative device and aesthetic eye-catcher with escapist dimensions—undulates between inner and outer spaces. Whether under water, in deep space or as an idea, the flying-floating gleaming jellyfish incites heightened experience of our environment. Simultaneously, it acts as a boundary crosser and boundary spanner between terrestrial and intergalactical spheres that points to our 'cosmic responsibilities': to the importance of understanding and carefully navigating our fragile environment—in and around ourselves and our planet. The cosmic jellyfish trope is associated with an expanded vision of what science can do, and creates scientific, cultural and individual meaning. It is a pulsating means for the juxtaposition and recognition of different cultural and scientific references, for the contrasting of ideas of the completely extraordinary and prosaic, of the supernatural and the environmental. Cosmic jellyfish spaceships in pop culture and fiction could cause NASA inferiority complexes, but fortunately they have already dreamed up their own jellyfish...

3 Jellyfish in Space Art: Transgressions and Non-Human Encounters with the Jelly Kind

Space jellyfish is a term introduced by NASA to describe an unusual atmospheric phenomenon resembling the shape of an UFO; in 2014 the Hubble Space Telescope found a strange looking jellyfish-like cluster of galaxies—now interpreted as a cloud of intergalactic dust in which many galaxies travel. From the beginning, space exploration was inextricably linked to technological progress and a special power of affirmation; it was supposed to serve the better understanding of human environments and an expansion of human consciousness.[22] Its dependency on technological progress makes it a phenomenon of the twentieth century, but it had precursors in other vertically oriented expeditions, such as deep-sea exploration or early atmospheric research using hot-air balloons. This is why the view into the oceanic depths and the view into space are historically and epistemically connected.[23]

In the 1980s, NASA initiated its "Artists in Space" programme to communicate scientific findings and the expansion of humans' spheres of life. In its young

[22] Cf. Malina, R.F (1991). In Defence of Space Art: The Role of the Artist in Space Exploration. *International Astronomical Union Colloquium. Light Pollution, Radio Interference and Space Debris*, 146, 145–152.

[23] Cf. Anzelewsky, N. (2017). Entdeckung und Eroberung des Meeres in der Moderne. München: Fink.

history, space art has been criticised for the commercialisation by corporations, Western-influenced research institutions' exploitative ideology and finally the ecological pollution of light and space. Above all, works of art that were left behind on other planets to "leave an imprint of (the artists') existence to be viewed by later civilisations" are now subject to postcolonial criticism.[24] This may be the reason why artists currently prefer ephemeral interventions or virtual media to engage with the cosmos. Nicola Triscott points out that "despite the more sceptical view of artists today concerning the rhetoric of the 1960s American space race, a lingering space of nostalgia and awe for the space programme remains".[25] This obscures the view of power-political questions of our time, such as space as a common good. In the UN Outer Space Treaty of 1967, the atmosphere, the deep sea and outer space were defined as "global commons", i.e. "it 'belongs' to all humanity".[26] This definition leads to questions of equality and ecological protection: "How can countries not already represented in a significant way in space take their place as the shared beneficiaries—culturally, technologically and economically—and trustees of this global commons?".[27] Contemporary space art attempts to find an answer to the ethical questions of space science. In order to raise ecological awareness, space art tests methods of how to address a wide, and possibly global, audience by evoking moments of intimate encounter / sensuous entanglement with the unearthly environment and its (imagined) extraterrestrial life forms, presented as submarine tentacular figures. In art, an environment which is extremely culturally charged because, or maybe although, it is only experienced by very few people, becomes communicated through its representation via a familiar creature of the submarine. Joanna Page has coined the term planetary art for artworks which "engender a more planetary perspective when it pays attention to forces we cannot compel".[28] Whatever the term, planetary or space art enables us to imagine cosmic environments as something familiar, which increases the sense of humans' connectedness with that which lies beyond the limits of our planet.

Researching the role of jellyfish in the arts means thinking about art and life in the future.[29] Jellyfish in the arts pose the question of what will thrive after humans have died out or left a world that they themselves have made uninhabitable for the human race.[30] These jellyfish are thus highly ambivalent phenomena—are they "survival species profiteers" or hopeful "prophets of a new epoch"?[31] As alien

[24] Cf. Malina, *In Defence of Space Art*, 150.

[25] Triscott, N. (2016). Transmissions from the Noosphere. Contemporary Art and Outer Space. *The Palgrave Handbook of Society, Culture and Outer Space*. Eds. J.S. Ormrod and P. Dickens, London: Palgrave, 424, 414–444.

[26] Ibid., 421.

[27] Ibid.

[28] Page, J. (2020). Planetary Art beyond the Human. Rethinking Agency in the Anthropocene. *The Anthropocene Review*, 7(3), 273, 273–294.

[29] Boetzkes A., Hiebert, T. (2022). *Artworks for Jellyfish and Others* (Introduction). Online: Nixious Sextor Press.

[30] Ibid., 3.

[31] Ibid., 4.

agents, figures of jellyfish stand on the threshold between science and arts, science and science fiction, knowledge production and the communication of scientific knowledge (Figs. 2 and 3).

On 18 June 2021, the "Neptune One" test spaceflight carried the artwork *Living Light*—a giant jellyfish designed by the *Beyond Earth* team—to above 99 percent of the Earth's atmosphere. A full-size capsule simulator had colour and light

Fig. 2 Beyond Earth: Living Light, 2021. © Beyond Earth

Fig. 3 Beyond Earth: Living Light, 2021. © Beyond Earth

shifts as it moved through space, powered by a space balloon.[32] The movement of spaceflight triggered colour and light shifts within the iridescent and fluorescent materials used in the design. Such changes that occurred—both internally and externally—not only made the vessel lifelike, but also alluded to the transformational experiences of spaceflight. At the peak of the giant jellyfish's journey, the integrated camera's look back offered a view of the curvature of Earth.[33] Here, a wondrous encounter with the fragility of our "Blue Marble" were combined with travelling on a cosmic jellyfish as an earthly-extraterrestrial experience. The intention of *Living Light* was to venture to a new world beyond Earth and return transformed, thus offering an altered perspective on our life on Earth and its location within a bigger whole. From this quest, the jellyfish—functioning as the non-human protagonist of the journey—returned as changed on its microbiological surface. *Living Light* is understood by the artist collective as a non-human explorational figure with the capability of transgression as well as connection: it brings earthly life to space and vice versa.

What is more, the luminous body of *Living Light* celebrates the oft-forgotten and unwitnessed beauty of the world, while also functioning as a testimony beyond the Earth's atmosphere and, maybe, existence. Launched from the coast, this sea-space creature sailed to low Earth orbit and landed in the Atlantic Ocean. The giant jellyfish was created as a seemingly autonomous being of artificial intelligence and thus expanded earthly life beyond our atmosphere. *Living Light* mimicked the bioluminescent and iridescent qualities of marine organisms and brought this light to outer space; to create a one-of-a-kind aquatic organism, *Beyond Earth* researched thousands of aquatic life forms to generate a dataset of biological structures that float, move and glow in the ocean. This is a scientific search for something like the first living being, comparable to the research of Johann Wolfgang von Goethe (who is considered one of the greatest German literary and scientist figures of the modern era) into the "Urpflanze" (a fictional plant archetype) in the eighteenth century. But the scientific and artistic collaboration went even further: the image dataset was used for machine learning to model the inner organ designs for *Living Light*, a method that combines biomimicry and artificial intelligence. Therefore, *Living Light* functioned as both a tentacular primordial being connecting the Earth's past with humans' present and future, and as a non-human agent which transgresses the earthly environment into outer space by fusing biological forms and technical innovation. As such, this space jellyfish stands on the threshold of being an artwork and creature; it is a material object with which humans can experience and reflect upon space and the Earth's place in it. As a figure of thought and a material object, the cosmic jellyfish opens up a diffuse contact zone: environmental connectedness is fostered by relational aesthetics and experiences.

---

[32] Cf. Beyond Earth, "Living Light", http://beyond-earth.org/projects-living-light/.

[33] Cf. Space Perspective, "Living Light View of Earth in Space", Facebook, https://www.facebook.com/watch/?v=1940711359438162.

This example from new artistic intervention in outer space shows that in the search of alternative ways of living and a way to expand humans' life to the realms of space, scientists turn to the deep sea for inspiration. This recent development is closely linked to scientific practice and funding policies. NASA funding includes some deep-sea explorations and astrobiologists have teamed up with marine biologists as well as oceanographers to test the boundaries of Earth-bound and unbound research (and filmmakers, see above). Exploring unknown dark territories of the deep sea, its creatures and their adaptation to the weakening of gravity to buoyancy (similar to zero gravity) reveals crucial similarities for exploring outer space. In its impenetrable-darkness, extreme pressure, coldness and disorienting equilibrium, deep sea creatures live in a world which is closer to outer space than Earth's land which make them such a fascinating object of scientific and artistic examination. This is especially true for jellyfish as tentacular entities which also function as a metaphor in environmental thinking. What is more, the tentacular becomes a leading figure within the conceptualisation and realisation of space experiments, testing how various aspects of what makes us human (including arts and culture) can be sustained in alien environments.

4 Jellyfish and Cosmic Responsibility: Conclusive Thoughts on the Environmental Message of Tentacular Creatures in Art and Popular Culture

"[T]he collaboration of space art and space science to sustain innovation is a critical factor in the success of off-world space preparation, and of maintaining new societies in space".[34] It is also a critical factor for the highly necessary reconsideration of human-nature-relations on Earth and beyond. Our future on Earth, and perhaps in space, may depend on it. Unpacking the aesthetic concepts, discourses and narratives associated with jellyfish in art and popular culture reveals the images of science and environmental exploration that emerge outside the framework; they extend into how we "explore and exploit the mirror images of science or scientists in the collective imagination".[35] Unwrapping the far out—in the form of the outer space jellyfish—underlines that our cultural and artistic expressions and media "are participants as well as producers of a dialogue about knowledge and have an important function within the public discourse".[36] In fact, visual fiction and art

[34] King, B. (2021). Space Art: The Cosmos. The Artist. The Invention of the Image. *ASCEND*, Nov.15–17, 1.

[35] Weingart, P., Hüppauf, B. (2007). *Science Images and Popular Images of the Sciences*. Eds. P. Weingart and B. Hüppauf, New York: Routledge, 6.

[36] Pansegrau, P. (2007). Stereotypes and Images of Scientists in Fiction Films. *Science Images and Popular Images of the Sciences*. Eds. B. Hüppauf and P. Weingart. New York: Taylor & Francis Group, 257.

Fig. 4 TA: *Technojellicus Cosmosis* 126 (2022). Orchestrated in MidJourney by TA

centred around the environment can open up "novel forms of seeing, of understanding interconnections"[37] by drawing attention to "the human enmeshment within the biosphere".[38] Given that environmental, scientific facts without cultural meanings and social significance are essentially meaningless and useless to society, understanding these entanglements, environmentality and enjoyment incarnated in cosmic jellyfish art, film and narratives may facilitate a deeper engagement with ecological science and environmental urgencies. This can contribute to a "healthy scientific culture within society"[39]—if not to a heightened sense of responsibility not only for Earth, but to a kind of cosmic consciousness for our planet and other possible homes for humans and other (non)living earthlings and spacelings in the universe (Fig. 4).

This chapter has uncovered the surprising ubiquity of the 'jellyfish in space' theme, which is pervasive in pop culture, space art and indeed 'techno art'. Australian artist TA's *Technojellicus Cosmosis* series (Figs. 1, 4 and 5), using state-of-the-art AI-powered software (MidJourney), is a striking example of technological jellyfish fantasies that transcend our world *and* traditional image production. TA's fantastically immersive technoid space creatures—digitally staged by artificial intelligence based on the artist's vision (expressed in keywords) and

[37] Löschnigg, M., Braunecker, M. 2019. Green Matters: Ecocultural Functions of Literature. *Nature, Culture and Literature 15*, Eds. M. Löschnigg and M. Braunecker. Leiden: Brill/Rodopi, 4.

[38] Westling, L. (2006). Literature, the Environment, and the Question of the Posthuman. *Nature in Literary and Cultural Studies: Transatlantic Conversations on Ecocriticism.* Eds. C. Gersdorf and S. Mayer. Amsterdam, New York: Rodopi, 25–47, 45.

[39] Burns, O'Connor and Stocklmayer. *Science Communication.* 197.

Fig. 5 TA: *Technojellicus Cosmosis* 507 (2022). Orchestrated in MidJourney by TA

visual strategies—are unsettling because of their ambiguous man-made or highly artificial origins. These alien figures seem to inhabit a dystopian world of their own, equally connected to human technoculture and biological life forms. The images speak of a post-apocalyptic yet memerising universe where the remnants of human culture have been transformed into something new—something that has its own creative power? (Fig. 5)

The tentacular and the idea of entanglement—including technological entanglement—thus go hand in hand and merge in the jellyfish creature and concept, which is in keeping with the spirit of Blue Humanities; but it also goes beyond it, for example when conveying environmental messages and highlighting the importance of science—if not science communication (see Cameron's approach to exploration). As we have seen above in many of our artistic and pop cultural examples, the tentacular jellyfish is a reflection on entanglement and the fear (and potentially hope) of possible escape as well as the threat of climate related catastrophes. The jellyfish in outer space is about both the idea of leaving a planet, ravaged by humans, to its own, and about the planet as something beautiful that needs to be protected. Precisely because jellyfish are in some respects formless creatures, they are transitory figures which can liquefy an established nature-human relationship and turn it into something new. Jellyfish are fluid figures of thought; they make moral questions about climate justice and climate awareness applicable to new environments and, through their concrete reference to sensual forms of life, give complex theories a concrete expression. Jellyfish fascinate and mesmerise; they make us aware of our connection to our Earth and—through the effortlessness with which they encounter us in artistic space environments—expand the human perspective to include the non-human. So, the final message could be: Don't be jelly!

Acknowledgments The authors would like to thank Canberra artist TA for creating the "Technojellicus Cosmosis" series especially for this book chapter.

Anna-Sophie Jürgens is a Lecturer (Assistant Professor) in Popular Entertainment Studies at the Australian National Centre for the Public Awareness of Science of the Australian National University (ANU) and the Head of the Popsicule, ANU's Science in Popular Culture and Entertainment Hub. Trained as a literary and cultural studies scholar, her research explores the cultural meanings of science, and science and comic performance in different media. Anna-Sophie has published on comic parasites and Joker viruses in (animated) fiction, environmental fragility in street art, and clowns and scientists in comics and popular theatre in numerous academic journals. Her recent books include "Circus and the Avant-Gardes" (co-editor, 2022) and "Circus, Science and Technology: Dramatising Innovation" (editor, 2020). Her next book, co-edited with Anne Hemkendreis, will investigate the power of popular art and aesthetics to communicate environmental urgency. Anna-Sophie is the editor-in-chief of the peer-reviewed open access online journal w/k—Between Science and Art (English section) and Associate Editor of the Journal of Science & Popular Culture.

Anne Hemkendreis is working as a Research Associate at the SFB 948 "Heroes—Heroisierungen—Heroizations" of the Albert-Ludwigs-Universität in Freiburg (Germany). She is also an Associate Senior Lecturer at the Humanities Research Centre of the Australian National University in Canberra and a member of the Young Academy of the National Academy of Sciences Leopoldina and the Berlin-Brandenburg Academies. Anne investigates the work of contemporary women artists who test heroic realms and visualise climate knowledge. Previously, Anne worked as a fellow at the Alfried Krupp Wissenschaftskolleg in Greifswald and as a research assistant at the Leuphana University in Lüneburg. Anne has taught at various universities, including the University of the Arts in Berlin, while performing on stage as a physical theatre and circus artist. She wrote her doctoral thesis on the visualisation of privacy in the paintings of the Danish painter Vilhelm Hammershøi (published in 2016). In 2023, her book "Communicating Ice through Popular Art and Aesthetics"—co-edited with Anna-Sophie Jürgens—will be published.

Space Travel: Human Cosmic Hitchhiker Concept

Christoffel Kotze

Abstract

The Covid-19 pandemic which started in 2020 is perhaps one of the most disruptive events in modern history from a socioeconomic point of view, it affected billions of people around the world. One could also argue that it took the topic of viruses as discussion point away from its specialist confines and transplanted it into the fore of the general discourse. In certain circles a notion is supported that lifeforms such as viruses could indeed have travelled to Earth by hitching a ride on a space object and impacting the surface by way of a meteorite (the theory of panspermia). In this article, the question is asked: what if human life were to spread through the universe from Earth in a similar way to ensure a continued existence and becoming a fully-fledged spacefaring species? The idea of this work is to take the concept of panspermia and suggest a way that intergenerational space travel can possibly be attempted by using spaceships, cosmic travelling objects and a viral approach. The topics will be explored in three sections, the first two dedicated to providing the necessary background information around the concepts involved such as viruses and panspermia. The third section will attempt at an interpretation of how human life could possibly spread through the cosmos through a what is called the Human Cosmic Hitchhiker concept.

C. Kotze (✉)
NOEZ Strategic Advisory, Cape Town, South Africa
e-mail: chris@noez.co.za

A. Froehlich (ed.), *Outer Space and Popular Culture*, Southern Space Studies,
https://doi.org/10.1007/978-3-031-25340-9_3

1 Introduction

> Your Worst Enemy Could Be Your Best Friend & Your Best Friend Your Worst Enemy"—Bob Marley.[1]

Covid-19 undoubtedly has become the most famous virus up and to the writing of this work. To put it into perspective—according to a Google search history analysis of 2020 the word 'Coronavirus' was the most searched for term in 2020 by a long margin.[2] Before 2020 it is highly unlikely that the name of a virus would have held the Google number one search spot, but in 2020 the topic of viruses was all of a sudden pushed to the front of day-to-day conversation for billions of people around the world by the arrival of a novel virus. This novel Corona virus, later to be named Covid-19,[3] was destined to become the cause of probably the largest socioeconomic upheaval in modern history when on 11 March 2020 the WHO declared the Covid-19 outbreak a global pandemic.[4] Traversing metaphoric uncharted waters, a number of very strict measures were suddenly introduced globally with the aim of rapidly curbing the spread of the disease, which at the time little was known about and with no vaccines readily available. From a social perspective most of the global population experienced these measures as a restriction on their freedom of movement like travel bans or confined to their place of residence for long periods and in certain countries even the purchase of certain products was prohibited. On the socioeconomic front, the impact of the pandemic and the measures to curb its spread were to have a colossal impact causing an estimated 7% of real global economic contraction, with the most severe impact unfortunately predicted for emerging economies.[5] In April 2021 the International Monetary Fund (IMF) in its World Economic Outlook Report,[6] predicted a significant economic 'scarring' resulting in a long-lasting decline in all the affected economic sectors, with high levels of future uncertainty due to the deep impact of the pandemic.

Also happening in 2021 was the 93rd awards ceremony hosted in Los Angeles by the Academy of Motion Picture Arts and Sciences where a short documentary became the first film from South Africa to triumph in its category.[7] *My Octopus*

---

[1] Quotepark, "Your Worst Enemy Could Be Your Best Friend & Your Best Friend Your Worst Enemy, https://quotepark.com/quotes/797389-bob-marley-your-worst-enemy-could-be-your-best-friend-your (all websites cited in this segment were last accessed and verified on 16 October 2022).

[2] Google, "See what was trending in 2020", https://trends.google.com/trends/yis/2020/GLOBAL.

[3] WHO, "Archived: WHO Timeline—COVID-19", 27 April 2020, https://www.who.int/news-room/detail/27-04-2020-who-timeline---covid-19.

[4] Ibid.

[5] Eduardo Levy Yeyati, Federico Filippini, 2021, *Pandemic divergence: A short note on COVID-19 and global income inequality*, Department of Economics Working Papers wp_gob_2021_06.rdf, (Buenos Aires: Universidad Torcuato Di Tella, 2021).

[6] International Monetary Fund, *World Economic Outlook: Managing Divergent Recoveries*, (Washington DC: IMF,2021).

[7] Jack Dutton, "My Octopus Teacher makes film history", 16 June 2021, https://newafricanmagazine.com/26202.

Teacher,[8] a short film compiled from footage shot over a period of time with the same octopus, in order to provide a unique perspective into her life, received the coveted golden Oscar statuette.[9] Naturalist diver Craig Foster started interacting with her during work documenting a new species of shrimp *Heteromysis octopodis* discovered just offshore of Cape Town in the False Bay area, unique as it maintains a symbiotic relationship with octopuses.[10] Foster incidentally also had one named after him for accidentally discovering during the same period *Heteromysis fosteri*—a shrimp which inhabits the empty shells of sea urchins and gastropods.[11] With a lifespan of eighteen months on the top end, *Octopus vulgaris,* the 'Common octopus', as does all species of octopuses has quite a number of very unique features when compared to other lifeforms.[12] Humans and, for that matter, most animals, birds and fish have the protein haemoglobin as oxygen carrier in their blood, the iron embedded within its structure responsible for the red colour. Octopuses as well as other members of the phylum Mollusca and certain spiders have blue blood, compliments of copper contained within its oxygen carrying protein hemocyanin.[13] A total of three hearts are involved with circulating the blue blood; two helping with oxygenation by pumping it around the gills with the third responsible for general circulation of the body.[14] Though not the only creature able to change skin colour for camouflage, they have the ability to do it very rapidly, as well as creating complicated patterns and modifying actual texture of the skin.[15] When they sleep, they change colour as well but in different ways, with some researchers speculating that it coincides with dreaming.[16] Having an individual brain for each of their eight tentacles give the 'arms' the ability to operate independently from each with an additional ninth donut shape central-brain, makes octopuses stand out in stark contrast to that of higher vertebrates with a centralised brain structure. Known for their intelligence, they have been documented to recognise different

[8] IMDb, "My Octopus Teacher", 2020, https://www.imdb.com/title/tt12888462.

[9] Matthew Carey, "Going Deep: How 'My Octopus Teacher' Became Both a High-Stakes Thriller and a Profound Commentary On Connection", 15 April, https://deadline.com/2021/04/my-octopus-teacher-craig-foster-feature-news-1234734693.

[10] Story Helen Swingler, " New shrimp species has unique association with octopus ", 15 August 2017, https://www.news.uct.ac.za/article/-2017-08-15-new-shrimp-species-has-unique-association-with-octopus.

[11] Ibid.

[12] MarineBio, "Common Octopuses, Octopus vulgaris", https://www.marinebio.org/species/common-octopuses/octopus-vulgaris.

[13] Beth Azar, Alisa Zapp Machalek, "Roses are red and so is... blood?", 14 February 2019, https://biobeat.nigms.nih.gov/2019/02/roses-are-red-and-so-is-blood.

[14] Lisa Hendry, "Octopuses keep surprising us - here are eight examples how", https://www.nhm.ac.uk/discover/octopuses-keep-surprising-us-here-are-eight-examples-how.html.

[15] Fox Meyer, "How Octopuses and Squids Change Color", https://ocean.si.edu/ocean-life/invertebrates/how-octopuses-and-squids-change-color.

[16] Donna Lu, "Octopuses may be able to dream and change colour when sleeping", 25 March 2021, https://www.newscientist.com/article/2272319-octopuses-may-be-able-to-dream-and-change-colour-when-sleeping.

people and reacted differently as well to individuals they seem to disapprove of. In many aquariums worldwide octopuses have been observed in activities that underscores their unique intelligence; 'play' with objects, escape their confines to steal food from an adjacent tank only to return to their own. A particularly interesting observation was made in New Zealand at the University of Otago where octopuses have learned to switch the overhead lamps of the aquarium off by squirting water at it and thereby destroying it with a short-circuit.[17] Research have shown that although octopuses expectedly react reflexively when confronted with a negative stimulus, they also quickly learn to actually actively avoid similar scenarios in the future.[18] A 2021 study confirmed that octopuses can not only feel pain but experience it as emotional distress similar to vertebrates even though there are radical differences in the architecture of the nervous system.[19] Last but not least, remarkably the octopus genome has a remarkable level of complexity having over 33.000 genes for protein-coding substantially more than the 20.500 found in the human genome.[20] With such an interesting and curious set of characteristics and abilities, it is no wonder that octopuses are often referred to as the closest thing to meeting an extra-terrestrial lifeform. In 2019 a research paper was published which proposed that the octopus might in actual fact very well be the product of the chance encounter between a primitive lifeform on Earth and an extra-terrestrial visitor in the form of a virus.[21]

A scenario where a virus from outer space comes to visit Earth has indeed been explored in the science fiction genre on several occasions, perhaps a good example for this work being the 1971 film, *The Andromeda Strain.*[22] In the film the population of a small town in New Mexico USA is almost completely wiped out after a satellite crashed not far from the town, the suspect being some unknown biological contaminant possibly of extra-terrestrial origin. Members of a specialist recovery team trained to deal with just such a scenario is sent in to investigate and contain the situation. The team determines that most of the towns people died rapidly as their blood literally crystallised, others simply went insane and committed suicide, there were however two sole survivors. It is subsequently confirmed that the satellite was indeed carrying an alien hitchhiker, a crystalline hydrocarbon-based life form quite dissimilar to the terrestrial organic based life forms. Named Andromeda, this virus-like organism is found to be highly virulent and as no one

[17] Radio New Zealand, "The secret life of the octopus", https://www.rnz.co.nz/national/programmes/afternoons/audio/2018641888/the-secret-life-of-the-octopus.

[18] Robyn J. Crook, "Behavioral and neurophysiological evidence suggests affective pain experience in octopus", *Iscience* 24, no. 3 (2021): 102,229.

[19] Carly Cassella, "Octopuses Not Only Feel Pain Physically, But Emotionally Too, First Study Finds", 5 March 2021, https://www.sciencealert.com/scientists-identify-the-first-strong-evidence-that-octopuses-likely-feel-pain.

[20] NIH, "Untangling the Octopus Genome ", 24 August 2015, https://www.nih.gov/news-events/nih-research-matters/untangling-octopus-genome.

[21] CORDIS, "Trending Science: Do octopuses come from outer space? ", https://cordis.europa.eu/article/id/123479-trending-science-do-octopuses-come-from-outer-space.

[22] IMDb, "The Andromeda Strain", https://www.imdb.com/title/tt0066769.

Table 1 What is an infection?

What is an infection?	
Source	Definition
Cambridge	"pass a disease to a person, animal, or plant"
Merriam-Webster	"contaminate with a disease-producing substance or agent (such as bacteria)"
Oxford learners	"make a disease or an illness spread to a person, an animal, or a plant"

can figure out how it works as it is so completely out of the range of conventional knowledge it amounts to an impossible task, though the team did manage to confine it at least. However, it did not take the alien organism long to escape its confines, surprisingly though instead of spreading more death and destruction it is now inert; its rapid mutation cycle has turned it into a harmless airborne organism at the time of its escape and it simply returns to space to continue its evolutionary journey of exploration.

The aim of this work is to look at a similar scenario, albeit in reverse, instead of an alien virus visiting Earth it will be humans leaving Earth spreading though space using a model based on the lifecycle of a virus. Consisting of three broad sections, firstly sketching a bit of background about the viral lifecycle, which will lead into a section exploring the concept of panspermia with the final section proposing a view of humanity as cosmic hitchhikers.

2 Viruses

The Covid-19 pandemic has established the word virus firmly in the lexicon of the average person, but what is a virus exactly? A virus can generally be described as a very small infectious parasite that operates at an intracellular level in biological organisms.[23] To understand the working of a virus it is necessary to firstly look at what is meant by the word 'Infect' and briefly what does the process entail? According to three widely used dictionaries of the English language (refer Table 1), there are slight differences in the definition of the word 'infection' however the word 'disease' occurs in every one of those definitions.[24,25,26]

The word disease in turn refers to a condition which is 'removed' from the normal expected condition of an organism, its etymology can be traced back to

[23] Hans R. Gelderblom, "Structure and Classification of Viruses", *Medical Microbiology*, 4th ed. (Galveston: University of Texas Medical Branch at Galveston, 1996).
[24] Merriam-Webster, "Infect", https://www.merriam-webster.com/dictionary/infect.
[25] Oxford Learners Dictionary, "Infect verb", https://www.oxfordlearnersdictionaries.com/definition/english/infect?q=infect.
[26] Cambridge, "Infect", https://dictionary.cambridge.org/dictionary/english/infect.

Table 2 Virus parasite

What is a Parasite?	
Source	Definition
Cambridge	"An animal or plant that lives on or in another animal or plant of a different type and feeds from it"
Merriam-Webster	"An organism living in, on, or with another organism in order to obtain nutrients, grow, or multiply often in a state that directly or indirectly harms the host"
Oxford Learners	"A small animal or plant that lives on or inside another animal or plant and gets its food from it"

an old French word '*desaise*' referring literally to a state of discomfort.[27] Another important definition to clarify is exactly what a parasite is (refer Table 2).[28,29,30]

For the purpose of this work, a parasite will refer to an organism that actively needs another to host its own survival in one way or the other—a relationship likely not beneficial to the host involved. A virus does certainly tick all the boxes to qualify as parasite, perhaps even the ultimate parasite, as a virus is entirely dependent on an external host to propagate its life cycle.

In biological terms a virus is fundamentally characterised by its inability to replicate itself without the presence of a suitable host. The disease effected by the virus in the infected organism is a direct cause of the process required for viral replication, the prime directive of the virus. In order to survive, a virus needs to do a couple of things right from the start, namely:

1. It needs to be able to survive while waiting to find a suitable host.
2. Secondly, upon infecting said host, it needs to avoid detection and possible destruction.

Once those two conditions are met, the virus can now get on with the business of replicating itself, thus generating a new cohort of viruses ready to repeat the process. For the virus to be as effective as possible in this task, it is advantageous to spread its copies as far and wide as possible, often deploying the services of 'third party' carriers to provide stealth. Take the example of a human being acting as host to a flu virus who is showing symptoms. Instinctively, another healthy human will strive to actively avoid this diseased person. Such a situation is not good for the virus as it needs to infect the healthy individual, but if the person does not enter the infectious sphere of the virus, it will not be able to do so

[27] Online Etymology Dictionary, "Disease", https://www.etymonline.com/word/disease.
[28] Cambridge, "Parasite", https://dictionary.cambridge.org/dictionary/english/parasite.
[29] Merriam-Webster, "Parasite noun", https://www.merriam-webster.com/dictionary/parasite?src=search-dict-box.
[30] Oxford Learners Dictionary,"Parasite noun", https://www.oxfordlearnersdictionaries.com/definition/english/parasite?q=parasite.

(incidentally the whole principle of quarantine). However, if the virus can manage to make things more difficult for its desired host to detect it, then the infection cycle will become much more effective—especially if the virus is hiding in another entity much smaller and much more pervasive (for example, yellow fever). Yellow fever is a haemorrhagic disease that has no cure—once a patient is infected, the morality rate for the unfortunate ones who develop a severe form of the disease can be as high as 60%. Caused by a RNA virus of genus *Flavivirus*, the carriers for the disease are mosquitoes, either *Aedes*[31] or *Haemagogus*[32] species. These mosquitos effectively take the virus to its preferred hosts; humans and primates. The infected host in turn also infects uninfected mosquitoes, turning them into even more carriers—a most effective cycle.[33] Yellow fever was a much-feared disease for many centuries with people erroneously believing it was contagious thus trying to avoid the diseased, without escaping the disease. This very successful cooperation between virus and mosquito was only discovered in 1900 when Reed and Carroll, two doctors in the USA, made the connection with the mosquito and coming up with the solution to kill the mosquitoes. By eliminating the carrier, you will eliminate the disease.[34]

Structurally a virus is relatively uncomplicated, constructed out of an outer coating layer consisting out of protein within which is encapsulated genetic material in the form of DNA or RNA. It is this genetic material the virus will use as 'blueprint' to make copies of itself using the resources of the infected host. The schematic (refer Fig. 1 Simplistic depiction of the viral lifecycle.) provides a very simplistic depiction of the viral lifecycle as a basic four phased cycle:

A. Infection takes place when a virus successfully penetrates the defences of a suitable host organism.
B. Using its genetic payload, the virus uses the host to start making many copies of itself.
C. Copies of the original virus now leaves the host which by this stage is resource deprived and in a state of disease. During this stage the viral genetic material may be altered due to environmental pressures resulting in the copy being different to the original. Known as a mutation, a new 'blue' variant of the same virus now joins the existing flock of its red kin.
D. A new cohort of viruses makes their way into the world looking for suitable hosts to infect to repeat the process.

[31] CDC, "Life Cycle of Aedes aegypti and Ae. albopictus Mosquitoes", https://www.cdc.gov/mosquitoes/about/life-cycles/aedes.html.
[32] Jáder da C. Cardoso, et al. "Yellow fever virus in Haemagogus leucocelaenus and Aedes serratus mosquitoes, Southern Brazil, 2008.", Emerging infectious diseases 16.12 (2010): 1918.
[33] CDC, "Transmission of Yellow Fever Virus", https://www.cdc.gov/yellowfever/transmission/index.html.
[34] America's Library, "U.S. Army Physicians Discovered the Cause of Yellow Fever. August 27, 1900", https://www.americaslibrary.gov/jb/progress/jb_progress_yellow_3.html.

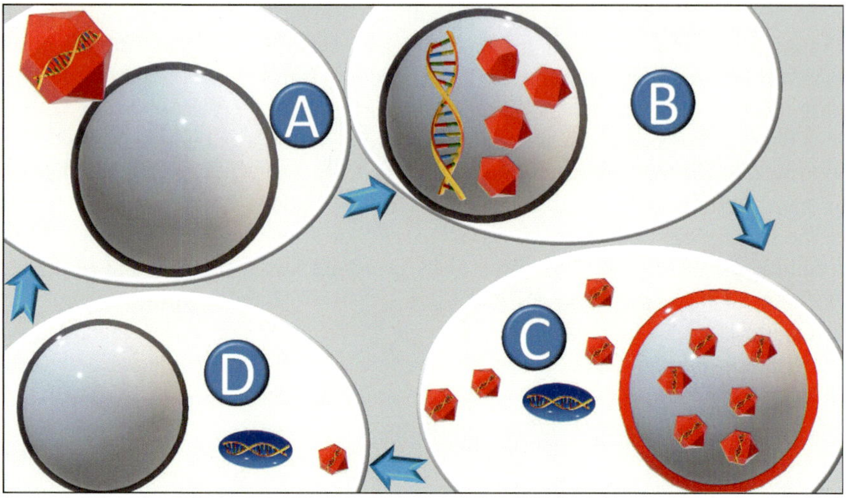

Fig. 1 Simplistic depiction of the viral lifecycle (Graphic courtesy of the author)

In this process of replication, the host organism can be severely damaged even to the extent that the host will cease to exist. This replication process can also alter the virus itself. Viral mutation is an important mechanism which the virus deploys to strengthen its future survival options, not dissimilar to the procreation of mammals where certain genes become favoured. Certain viruses have a better ability to mutate than others. It is an important feature enabling the virus to adapt to new environments and even different hosts. Referred to as zoonotic spill-over some viruses mutating from the original now will have the ability to not only infect their original host species, say simians for arguments sake but now also humans. The Ebola virus family is suspected of such a "jump" between species.[35] It is believed that many factors relating not only to the virus but also to the host, are at play driving the production of almost brand-new virus versions in a short period of time.[36] The Covid-19 pandemic saw a fairly quick production of mutations with different versions of the virus originating in different parts of the world as early as September 2020, with a variance in effect of the infections as well.[37] Not all virus mutations will survive, but if one seems to be particularly successful for a set of circumstances it will ultimately become the dominant version of the virus.

[35] Alla Katsnelson, "How do viruses leap from animals to people and spark pandemics? ", 30 August 2022, *Chemical and Engineering News*, Volume 98, Issue 33, https://cen.acs.org/biological-chemistry/infectious-disease/How-do-viruses-leap-from-animals-to-people-and-spark-pandemics/98/i33.

[36] Rafael Sanjuán, Pilar Domingo-Calap "Mechanisms of viral mutation", Cellular and molecular life sciences 73.23 (2016): 4433–4448.

[37] WHO, "Tracking SARS-CoV-2 variants", https://www.who.int/activities/tracking-SARS-CoV-2-variants.

At the end of the day, the purpose of a virus infecting the host is thus not one of wilful destruction but very much a matter of survival when viewed from the point of view of the virus. One can further argue that a virus is in fact neither generically "good nor bad" it entirely depends on the perception of the observer and the situation. Take the following example: in 1917, before the discovery of antibiotics, Félix d'Hérelle (a French microbiologist) used a solution with what he called bacteriophages (a virus that only kills bacteria) to cure dysentery successfully. Interestingly, the use of these bacteriophage as a treatment agent is again finding new applications in modern medicine.[38] In this case, the virus simply did what it was destined to do—it used the bacteria as host to copy itself, in the process the bacteria die, so from the point of view of the bacteria the virus most certainly is a killer. The perspective from the human patient being cured by virtue of the virus taking out the disease-causing bacterial pathogen, is expectedly one of gratitude.

At the end of the day, it is more than likely that the general perception of viruses and their often-devastating effects on society (directly or indirectly) are justifiably perceived as negative. However, there is much more that can be learned from these smallest of all perceived enemies. Perhaps the most obvious thing that one can learn from it is its apparent tenacity, once it has set its course it proceeds with a ruthless effective resilience in completion of its mission, rapidly adapting itself to ever changing conditions.

At this point the reader can quite rightly ask the question: what does the lifecycle of a virus have to do with space travel? Before attempting to answer that question, it is pertinent to introduce another, perhaps not that well-known a concept at this point, which will be that of panspermia. Derived from an ancient Greek word *panspermos*[39] which literally meant 'containing all kinds of seed' the concept of panspermia concerns itself with the origins of life on Earth, and, for that matter, life in the universe. The next section will briefly explore this very complex and controversial concept in an attempt to provide the background necessary in context of this work.

3 Panspermia

Dandelions are very common plants of which variants are found all over the world. They propagate by having their seeds float away with the ambient wind and if it falls back to Earth at a suitable location, another plant will grow. Panspermia can be seen as the cosmic equivalent of a dandelion—one where the 'seeds' of life are spread throughout the universe using asteroids, comets, and other interstellar objects as transport. The theory has been around for thousands of years: first traced back to a philosopher in ancient Greece around 500 Before the Common

[38] Stephanie Dutchen, "The Good that Viruses Do", Viral World, Spring 2022, https://hms.harvard.edu/magazine/viral-world/good-viruses-do.

[39] LEXICO, "Panspermia", https://www.lexico.com/definition/panspermia.

Era (BCE) but the concept it seems is not well known amongst the general population.[40] In a very brief informal social media questionnaire done by the author, distributed amongst peers, less than 8% of the respondents answered, 'yes' to the following question "Have you ever heard of the term panspermia?" This section will try to illuminate the subject by providing a brief overview.

On the moon there is a prominent impact crater located not far from the Lunar north pole called *Anaxagoras* (named to honour the philosopher Anaxagoras).[41] Credited as the first person to correctly explain the true mechanisms behind the nature of an eclipse, having surmised correctly that the Moon was a 'rock' and the Sun a 'burning rock.' Sadly for Anaxagoras this theory at the time did not lead to accolades but rather arrest and banishment from Athens.[42] Born in what is modern day Turkey approximately 500 BCE Anaxagoras is noted as the first of the Presocratic philosophers to be based in Athens and one of three Pluralists.[43] Pluralism, from a cosmological perspective, views the universe as consisting out of not one or two but rather a multitude of basic building blocks.[44] In the view of Anaxagoras, the original universe consisted out of many different 'ingredients' existing in a thoroughly mixed state as a collective of no particular distinction, which at some point morphed into something more distinctly identifiable; through the application of a divine *nous*.[45] Generally associated with the quote "all things are in everything"[46] Anaxagoras is also seen as the originator of the concept of Panspermia (literally meaning life is everywhere) and as such, the 'seeds' of life can traverse the universe to express itself where suitable.[47]

Since its ancient origins, the concept has been revisited from time to time only to mostly be rejected by the ruling intelligentsia of the time and relegated back to historic curiosity waiting to be resurrected yet again at some future point. In the eighteenth century French author Benoît de Maillet, an employee of the French Court, while based in Egypt took a keen interest in studying the history of the

---

[40] Temple, Robert., "The prehistory of panspermia: astrophysical or metaphysical?", International Journal of Astrobiology 6, no. 2 (2007): 169–180.

[41] Veronica Bray, "Exposed Fractured Bedrock in the Central Peak of the Anaxagoras Crater", 14 April 2010, https://www.lroc.asu.edu/posts/155.

[42] David Warmflash, "An Ancient Greek Philosopher Was Exiled for Claiming the Moon Was a Rock, Not a God", 20 June 2019, https://www.smithsonianmag.com/science-nature/ancient-greek-philosopher-was-exiled-claiming-moon-was-rock-not-god-180972447.

[43] Patricia Curd, "Anaxagoras", 11 November 2019, https://plato.stanford.edu/entries/anaxagoras/#EveEve.

[44] Philosophy Basics, "By Movement / School > Ancient > Pluralism", https://www.philosophybasics.com/movements_pluralism.html.

[45] "Patricia Curd, "Presocratic Philosophy", 22 June 2020, https://plato.stanford.edu/entries/presocratics/#PluAnaClaEmpAcr.

[46] John Palmer, "Everything in Everything: Anaxagoras's Metaphysics", 7 November 2017, https://ndpr.nd.edu/reviews/everything-in-everything-anaxagorass-metaphysics.

[47] Yuko Kawaguchi, "Panspermia Hypothesis: History of a Hypothesis and a Review of the Past, Present, and Future Planned Missions to Test This Hypothesis", 2019, https://ui.adsabs.harvard.edu/abs/2019asbi.book..419K/abstract.

ancient world. During this process, the concept of panspermia also somehow got revived again. He speculated about the origins of life in the book *Telliamed* in which a number of radical ideas were put forward about the origins of Earth. These ideas were so radical that the book was only officially published a decade after de Maillet's own demise in 1748.[48] Versions of the book had previously been available for many years unofficially but due to the controversial nature of its content, it was heavily censored. The nineteenth century saw many scientific discoveries happening simultaneously—allowing new questions to be asked about many things including the origins of life. The concept of panspermia was unsurprisingly brought to the fore again. In this era of discovery, panspermia was not only revived but actively supported by some of the great scientific luminaries of the time such as Lord Kelvin and Hermann von Helmholtz amongst others.[49]

In 1908 Svante Arrhenius, a Nobel laureate from Sweden, added additional credence to the theory by suggesting that not only is panspermia possible but actually suggested a mechanism of its workings; it was radiation from the sun driving bacterial spores, the propagators of life, through space.[50] Though this was a big win for pro-panspermia pundits, they were to receive a harsh blow to their argument when Paul Becquerel proved that bacterial spores would more than likely not survive the rigours of exposed space travel due to vacuum, extreme temperatures and radiation.[51] In the 1920s, Soviet biochemist Alexander Oparin presented the idea that that life on Earth developed from the evolution of carbon-based molecules from a "primordial soup"—essentially life originating from 'non-life' or abiogenesis.[52] This idea presented an alternative and increasingly plausible view, when through a series of later experiments in the 1950s, amino acids were indeed synthesized from a feedstock of chemicals with the addition of electricity; the detractors of panspermia took note. Since then, a lot has happened. Several experiments conducted in space with the advent of the new millennium have started to provide new insight into the ability of biological life to survive the rigours of space.

In a 2005 European Space Agency (ESA) sponsored experiment where different microorganisms were exposed to space for sixteen days. It was found that while generally most bacteria were killed by exposure to hostile space conditions such as cosmic rays and massive UV exposure, certain lichens could indeed survive.[53] A

---

48 Michon Scott, "Benoît de Maillet", 25 June 2022, https://www.strangescience.net/demaillet.htm.

49 New Scientist, "The word: Panspermia", 1 March 2006, https://www.newscientist.com/article/mg18925411-900-the-word-panspermia.

50 Yuko Kawaguchi, "Panspermia hypothesis: history of a hypothesis and a review of the past, present, and future planned missions to test this hypothesis." Astrobiology (2019): 419–428.

51 Stephen Fleischfresser, "A brief history of panspermia The ancient history of a surprisingly resilient idea", 23 April 2018, https://cosmosmagazine.com/nature/evolution/over-our-heads-a-brief-history-of-panspermia/.

52 The Physics of the Universe, "Alexander Oparin (1894–1980)", https://www.physicsoftheuniverse.com/scientists_oparin.html.

53 Leopoldo G. Sancho, et al. "Lichens survive in space: results from the 2005 LICHENS experiment", Astrobiology 7.3 (2007): 443–454.

subsequent 2015 experiment onboard the EXPOSE-E facility of the International Space Station (ISS) showed that lichens could survive even a much longer exposure to space conditions. A fraction of the original experiment sample of *Xanthoria elegans* survived for more than a year and a half.[54] But it has not only been bacteria and lichens that have stood the space test. In 2007, an ESA experiment exposed about three thousand Tardigrades to space onboard the ESA FOTON-M3—an ISS experiment for twelve days, the creatures survived without a problem.[55] Tardigrades colloquially known as 'water bears' are found all over the Earth from the extreme pressure of the deepest oceans to high-up in the mountains where the air is thin, in temperatures ranging from equatorial heat to the freezing conditions of the arctic.[56] These extremely hardy micro animals have been proven to be able to withstand such punishing conditions surviving not only extreme terrestrial conditions, but even the vacuum and radiation of space.[57]

It is therefore not surprising that the theory of panspermia is finding a receptive audience in an era of modern science where space exploration is providing the platform to test aspects of the theory. In their paper *Cause of Cambrian Explosion—Terrestrial or Cosmic?*[58] the authors (Steele et. al) present their case for panspermia believing their data provides a strong case in support of panspermia. Written by a multidisciplinary group of thirty scientists, using various studies over more than sixty years on the topic, the authors believe they make a compelling case for panspermia. Though the paper has been criticized as being biased and has attracted a fair amount of criticism, the biological experiments done in space does indicate that certain lifeforms do indeed have the ability to survive the rigours of space under certain conditions. The question can thus rightly be asked—why do these different organisms have the ability to survive space exposure in the first place? What evolutionary driver could possibly be involved?

Politicians in established systems of governance tend to crave stability and thus anything that might potentially disrupt the status quo is, not unsurprisingly, typically rejected and rather left for a future government or power structure to deal with. Strong positions on scientific viewpoints can also take a long time to change, especially ones touching on fundamental issues such as the origin of life. These concepts not only have its set of vested academic interests but also due to deeply

[54] Annette Brandt, et al., "Characterisation of growth and ultrastructural effects of the Xanthoria elegans photobiont after 1.5 years of space exposure on the International Space Station", Origins of Life and Evolution of Biospheres 46.2 (2016): 311–321.

[55] ESA, "Tiny animals survive exposure to space", 10 November 2008, https://www.esa.int/Science_Exploration/Human_and_Robotic_Exploration/Research/Tiny_animals_survive_exposure_to_space.

[56] Sarah Bordenstein, "Tardigrades (Water Bears)", https://serc.carleton.edu/microbelife/topics/tardigrade/index.html.

[57] Alina Bradford, Mindy Weisberger, "Facts about tardigrades", 12 November 2021, https://www.livescience.com/57985-tardigrade-facts.html.

[58] Edward J. Steele, et al., "Cause of Cambrian explosion-terrestrial or cosmic?", Progress in Biophysics and Molecular Biology 136 (2018): 3–23.

integrated socioreligious aspects in many societies, it has potentially a disruptive political dimension. One only needs to think back that it took more than a thousand years for the original Ptolemaic geocentric model to be challenged successfully by heliocentric theory.[59] Published by Nicolaus Copernicus in 1543 of the Common Era (CE) it suggested the sun was indeed the centre around which all planets rotated. It was not well received and in fact the book was banned by the de facto political force at the time, the Catholic Church in 1616.[60] Only when Johannes Kepler published 'Kepler's Three Laws' of celestial mechanics in 1619 CE which was based upon the principle that the Sun was indeed the centre of the solar system, did a gradual acceptance start to take place.[61,62]

Panspermia can be seen as a process involving a 'triple survival event' one which the travelling piece of 'life-seed' needs survive before calling its quest a success. The organism or piece of genetic material involved firstly needs to escape its current environment, for example a planet intact, next it needs to be able survive its travel in space and finally it needs to find a place to land where it can thrive.[63] For the purposes of this work the following terms will be used to refer to the triple survival event:

A. Donor Object Escape (DOE)—the life form finds a way to break free from the boundaries of the object where it was able to survive. There are basically two ways that is generally looked at to trigger such an event namely ballistic panspermia and directed panspermia.
   a. Ballistic panspermia—a collision event between objects with at least one acting as donor of the life-seeding material, the collision energy providing the necessary escape velocity.[64] Note that Lithopanspermia is a term used in conjunction with ballistic panspermia, sometimes interchangeably. In such a case the life-seeding mechanism is encased within the bit of debris dislodged by the collision, providing the necessary protection against the harsh space environment.[65]

---

[59] Sheila Rabin, "Nicolaus Copernicus", *The Stanford Encyclopedia of Philosophy* Fall 2019 Edition, Edward N. Zalta (ed.), https://plato.stanford.edu/archives/fall2019/entries/copernicus.

[60] Maurice A. Finocchiaro, "400 Years Ago the Catholic Church Prohibited Copernicanism", 2019, https://origins.osu.edu/milestones/february-2016-400-years-ago-catholic-church-prohibited-copernicanism?language_content_entity=en.

[61] Daniel A. Di Liscia. "Johannes Kepler." Edited by Edward N. Zalta. *The Stanford Encyclopedia of Philosophy*, no. Fall 2019 Edition (2019).

[62] Jack J Lissauer, "In Retrospect: Kepler's Astronomia Nova", 9 December 2009, https://www.nature.com/articles/462725a.

[63] Yuko Kawaguchi, "Panspermia hypothesis: history of a hypothesis and a review of the past, present, and future planned missions to test this hypothesis", Astrobiology (2019): 419–428..

[64] Geological Digressions, "The origin of life: Panspermia, meteorites, and a bit of luck", https://www.geological-digressions.com/the-origin-of-life-panspermia-meteorites/

[65] Pabulo Henrique Rampelotto, "Panspermia: A promising field of research", In Astrobiology Science Conference 2010: Evolution and Life: Surviving Catastrophes and Extremes on Earth and Beyond, vol. 1538, p. 5224. 2010.

b. Directed panspermia—as the name implies this a wilful attempt at seeding life throughout space.[66]

B. Space Wandering Phase (SWP)—the life from is travelling through space until a point of intervention stops this phase, for example a collision with another space object. There are basically two modes of what could 'propel' the object on during its pace journey.
   a. Kinetic energy it acquired as a result from the ballistic event that led to its escape will propel it onwards until it ultimately finds an object to collide with / land on.
   b. Radiopanspermia—small individual space hardened lifeforms present in space can be propelled towards a collision event using only pressure provided by ambient space radiation.[67] Panspermia as process does not conceptually differentiate per se between interplanetary and interstellar and in actual fact can be both at any given point, depending on perspective.
C. Recipient Object Touchdown (ROT)—due to an intervention event, typically in the form of a collision or through radiopanspermia, the life-seeding material successfully performs a touchdown on a recipient object where it may be able to not only survive but thrive.

These events should not be seen as strictly discreet events but in fact there can potentially be significant overlap. Figure 2 presents several scenarios to illustrate some of the key concepts mentioned.

A. In the first scenario an object containing life-seeding material collides with planet Earth representing a ROT as it deposits its life-seeding payload. It also causes a subsequent DOE as the impact has such a velocity that pieces of Earth escapes to space to enter the SWP until it finds an event to end its journey.
B. A representation of radiopanspermia, in this scenario bacteria and viruses able to survive the rigours of space are propelled by radiation from a star.
C. Both products of radiopanspermia and ballistic panspermia roaming space in search of an event to stop its journey.
D. Graphic representation of a lithopanspermia object, the life seeding material encapsulated within a piece of debris dislodged from the donor object now traveling through space in search of a suitable recipient.
E. The ROT marks the end of the journey, the Life-seeding objects encounter a planet suitable for the development of life or indeed where existing life can be altered. As mentioned under point A this phase can indeed also initiate the start of a new cycle of panspermia again.

---

[66] Oskari. Sivula, "The Cosmic Significance of Directed Panspermia: Should Humanity Spread Life to Other Solar Systems?", Utilitas (2022): 1–17.

[67] Jeff Secker, Paul S. Wesson, James R. Lepock, "Astrophysical and biological constraints on radiopanspermia", arXiv preprint astro-ph/9607139 (1996).

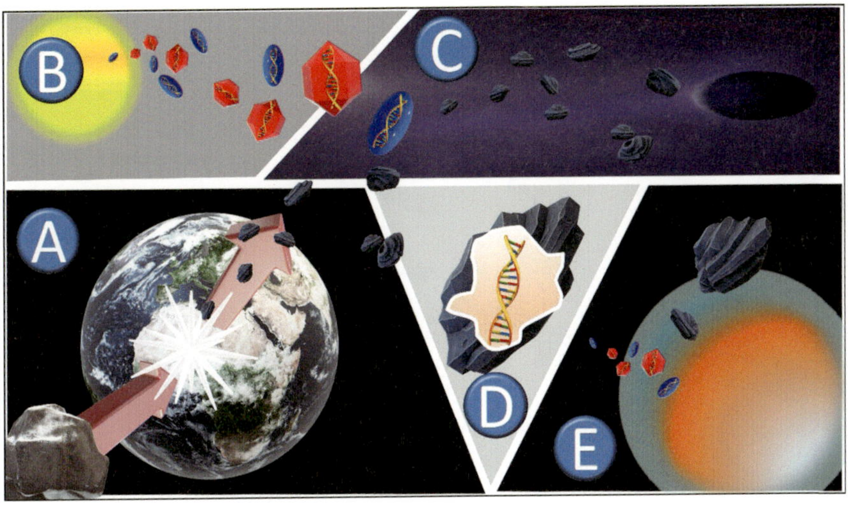

Fig. 2 Panspermia—a graphic representation (Graphic courtesy of the author)

As the focus point of this work will be ballistic panspermia, the next section will look at it in a bit more detail.

3.1 Ballistic Panspermia

In applied physics, ballistics is the area that concern itself with the motion of an object which does not propel itself, but rather acquires its motion as the result of an initialisation event. It generally can be seen as having three areas of interest: the launch event, flight event and termination event—very analogous to the three phases of panspermia as previously discussed. To demonstrate the basic concept, take the typical example of a non-explosive projectile being fired from a cannon towards the target, a distant mountain on a firing range as depicted in Fig. 3.

A. Launch event, during this event a rapid chemical reaction of a propellent causes a sudden expansion of gas forcing the projectile to exit the barrel at high-speed.
B. Flight event, the moment the projectile is set into motion it enters its flight phase, during which it will be subjected to a variety of forces that will influence the event, for example, gravity, air density, wind speed and the likes.
C. Termination event, as the name suggests during this phase the motion of the projectile is suddenly terminated as it impacts the mountain. Both the projectile and the target at this point experience tremendous forces acting on it, as the kinetic energy is converted into deformation and/or transferred to other objects such as pieces of the mountain becoming projectiles themselves and thus the termination event sets in motion another launching event as debris.

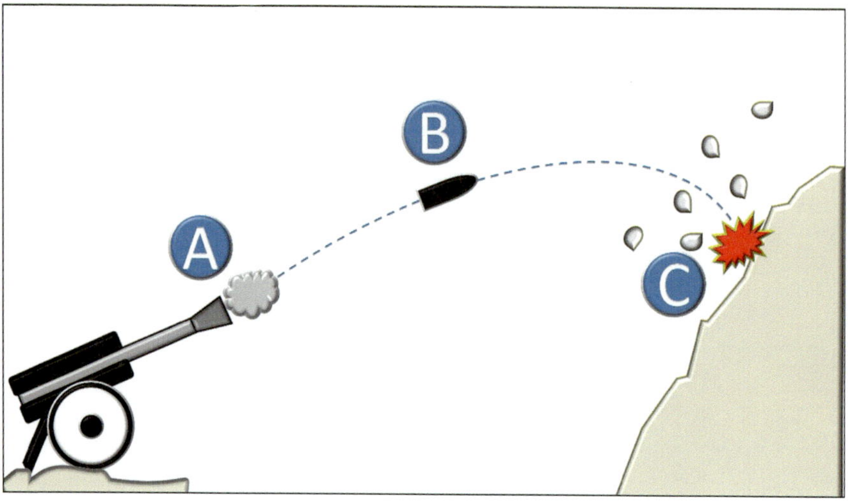

Fig. 3 Basics graphic of representation of ballistics (Graphic courtesy of the author)

Objects travelling in space typically move at great speed and when it encounters another object, and a collision ensues, it is normally referred to as a 'hypervelocity' impact event. Such a hypervelocity impact event is defined by the European Space Agency as one where the relative velocity exceeds 4.000 m/s and can in fact be much higher (a fast meteor can clock speeds exceeding 250.000 kmph).[68,69] Ballistic panspermia thus theorises that when a space object, such as a meteorite, impacts a planet like Earth the resulting hypervelocity event could transfer sufficient kinetic energy to enable debris created by said impact to exit the gravitational pull of the planet and enter a trajectory in space. At this point it might be pertinent to have a look at some important definitions of space objects in its different forms which are typically associated with ballistic panspermia, asteroids, comets, and meteoroids.

4 Asteroids

Asteroids are essentially atmosphere-less space 'space rocks' that can take a variety of orbits, ranging in size from a couple of meters in length, or as is the case of asteroid *Vesta,* to over 530 km long. They can also vary greatly in composition and

[68] ESA, "What are hypervelocity impacts?", www.esa.int/Enabling_Support/Operations/What_are_hypervelocity_impacts.

[69] National Geographic Society, "Meteor", https://education.nationalgeographic.org/resource/meteor.

shape.[70] Asteroids are rarely spherical in shape, lacking the gravitational 'muscle' to attain such a shape rather most of them present as random irregular shapes most often displaying collision damage in the form of impact craters. Asteroids rotate whilst in their solar orbit and also around each other as pairs; as much as 15% of all asteroids are believed to be in such a binary or even triple configuration.[71] Asteroids are generally classified in terms of their orbit type and secondly in terms of composition.

Orbit wise, most asteroids in the solar system can be found clustered between the orbits of Mars and Jupiter in an area known as the asteroid-belt, a ringed formation resembling a doughnut shape.[72] Apart from the asteroid-belt, they can also be found at the various Lagrange[73] points of planets, an object such as an asteroid finding itself at this point will experience an equilibrium of gravitational pull between the planet and the Sun rendering it stationary. These asteroids are known as Trojans and will thus follow the orbit of the planet it is 'bound' to.[74] Another orbit where asteroids can be found and one more significant to Earth are those classified as Near-Earth objects (NEOs) or objects with orbits close to that of Earth thus creating a potential collision hazard. Currently 20,000 asteroids are classified as NEOs.[75]

As far as composition is concerned it is not an easy task to accurately determine from a distance what exactly asteroids are composed of just by observing them. General speculation of what constitutes asteroids can be made by studying chunks of asteroids ending up on earth as meteorites. This knowledge can thus be combined with spectral analysis of asteroids to make certain basic deductions about their construction. Asteroids can also be classified into many subclasses in terms of their spectral signatures for which a number of taxonomy systems have been developed, for example those by Bus-DeMeo[76] and Tholen[77] which are helpful to distinguish different groupings of asteroids from afar. In general, though asteroids

[70] Alice Fuller, "These are the biggest asteroids and the threats they pose to Earth ", 25 October 2021, https://nypost.com/2021/10/25/these-are-the-biggest-asteroids-and-the-threats-they-pose-to-earth.

[71] ESA, "Binary asteroids, key to Earth's planetary defence", 2019, https://www.esa.int/Space_Safety/Hera/Binary_asteroids_key_to_Earth_s_planetary_defence.

[72] NASA Solar System Exploration, "10 Things: What's That Space Rock?", 21 July 2021, https://solarsystem.nasa.gov/news/715/10-things-whats-that-space-rock.

[73] Holly Spanner, "What are Lagrange points?", 11 February 2022, https://www.sciencefocus.com/space/lagrange-points.

[74] Charles Q. Choi, Ailsa Harvey,"Asteroids: Fun facts and information about these space rocks", 22 November 2021, https://www.space.com/51-asteroids-formation-discovery-and-exploration.html.

[75] ESA, "Near-Earth Objects - NEO Segment", https://www.esa.int/Space_Safety/Near-Earth_Objects_-_NEO_Segment.

[76] Planetary Science Institute, "Bus-DeMeo Asteroid Taxonomy", 14 August 2019, https://sbn.psi.edu/pds/resource/busdemeotax.html.

[77] David J. Tholen, *Asteroid taxonomic classifications,* Asteroids II, 1989.

Table 3 Chemical Classification of Asteroids

Chemical classification of asteroids	
Type	Description
C	Also known as carbonaceous asteroids, they contain some metals and silicates, but are mostly made out of carbon with a very low albedo of 0.03 to 0.10, absorbing virtually all sunlight and making them very hard to detect. Representing about 75% of all asteroids, they are found on the outer edge of the asteroid belt between Mars and Jupiter and are believed to be the oldest rocks in the solar system
S	Also known as silicaceous asteroids, they are made up of mostly silicate and iron-nickel and they are the second most abundant form of asteroid. They mostly inhabit the inner edge region of the asteroid-belt and represent about 17% of the total asteroid tally. An albedo of 0.2 makes them much easier to observe than their C-Type counterparts by reflecting much more sunlight
M	The metallic asteroids are the least abundant form of asteroid, basically making up the remaining 8% of the asteroid population and are mostly located in the central part of the asteroid-belt. Most are believed to contain iron and nickel, but less is known about any additional components to their composition than is the case with the C and S-Types. It is believed that their structure will change the closer they are to the Sun where a more molten appearance will more likely be the case

are classified in terms of basic chemical composition as one of three types either C, S or M refer Table 3.[78,79]

There is another way to study asteroids and that is to send a spacecraft to study it up close using special instrument packs. Several such missions have successfully been contributing to the body of knowledge of asteroids on earth refer Table 4[80] with a number of future ones planned as well. The ultimate prize is to collect a piece of asteroid and bring it back to Earth for study—as of the writing of this work in 2022, two such missions have been successful. In 2010 Hayabusa, a spacecraft from the Japanese Aerospace Exploration Agency (JAXA), became the first craft to not only collect a sample from an asteroid but to return it to Earth; a very small piece was obtained from *25,143 Itokawa* (a near-Earth asteroid). Though the spacecraft broke up on 13 June 2010 during re-entry, the sample capsule survived the touchdown in Australia and was successfully recovered.[81]

In December 2020, another JAXA spacecraft, Hayabusa 2, collected and delivered a much larger sample of *Ryugu* another near-Earth asteroid after a mission

78 NASA NEAR, "Near Earth Asteroid Rendezvous (NEAR) Press Kit", February 1996, https://www.nasa.gov/home/hqnews/presskit/1996/NEAR_Press_Kit/NEARpk.txt.

79 Mark Smit, "Comets vs asteroids: How do these rocky objects compare?", Space.com, 23 May 2022, https://www.space.com/comets-vs-asteroids.

80 NASA Space Science Data Coordinated Archive, "Asteroids ", https://nssdc.gsfc.nasa.gov/planetary/planets/asteroidpage.html.

81 Elizabeth Howell, "Hayabusa: Troubled Sample-Return Mission", 31 March 2018, https://www.space.com/40156-hayabusa.html.

Table 4 Asteroids related space missions, past present and future

Space missions contributing to the study of asteroids			
Craft	Mission type	Asteroids visited	Launched
Cassini	NASA/ESA Mission to Saturn	Through Asteroid Belt	1997
Deep Space 1	NASA Flyby Mission	Braille	1998
Rosetta	ESA Comet Mission fly by	Steins and Lutetia	2004
Dawn	NASA Orbiter	Ceres and Vesta	2007
DART	Kinetic Impact	Didymos and Dimorphos	2021
Lucy	Flyby Mission	Multiple Trojan Asteroids	2021
NEA Scout	Flyby CubeSat Mission	Near Earth Asteroid	2021
Psyche	Orbital Mission	Main Belt Asteroid 16 Psyche	2022
Hera	Follow-up Mission	Didymos and Dimorphos	2024

lasting just over six years.[82] The NASA spacecraft OSIRIS-Rex successfully collected a sample from asteroid *Bennu* in October 2020, which it will send to Earth when it passes by on 24 September 2023, after which it will redirect to intercept asteroid *Apophis*, which will be in the vicinity for an 18-month data collection study renamed as mission OSIRIS-APEX.[83]

5 Comets

Comets have a distinct appearance characterised by a glowing ball (the nucleus) and a trailing tail which could be millions of kilometres long. Their solar orbit

[82] Meghan Bartels, "Japan's asteroid samples faced surprise challenges on Earth: A pandemic, traffic jams and airport security", 29 April 2022, https://www.space.com/hayabusa2-asteroid-samples-japan-nasa-travel-challenges.

[83] Mikayla Mace Kelley, "NASA gives green light for OSIRIS-REx spacecraft to visit another asteroid", 25 April 2022, https://www.asteroidmission.org/?latest-news=nasa-gives-green-light-for-osiris-rex-spacecraft-to-visit-another-asteroid.

period can range from less than 200 to over 250.000 years on the other extreme.[84] Unlike asteroids it is not a solid object per se, rather its core is mostly made up of ice and some frozen gas which can also incorporate dust and bits of rocks.[85] Depending on the position of the comet relative to the sun, the size of the nucleus will change. Far away from the sun the nucleus in a frozen state can vary in size on average about 16 km across, though larger comets have been observed.[86] In 2022 the Hubble space telescope picked up what was the largest nucleus observed at the time, comet *C/2014 UN271 (Bernardinelli-Bernstein)*—it is believed to over 128 km in diameter and weighing an estimated 500 trillion tonnes.[87] As a comet approaches the sun the nucleus will expand to a size much bigger than most planets in the solar system due to the sun's heating effect.

6 Meteoroids, Meteors and Meteorites

Some of the most common objects in our solar system and perhaps most applicable to ballistic panspermia are meteoroids, meteors, and meteorites. These three different terms basically describe the same object. Surprisingly the fact that meteorites came from space was only acknowledged scientifically in 1803 after a mass meteorite event in L'Aigle in Normandy.[88] French physicist Jean-Baptise Biot was able to convincingly prove similarities between the 'stones' recovered from the L'Aigle incident with that of a similar event which occurred in the town of Barbotan more than a decade before.[89] In 2021 one of the Barbotan meteorites, one with a macabre twist as it had reputedly killed a person inside a hut when it fell through the roof, was sold on auction fetching almost USD 3.000.[90] It is important to distinguish between these terms, although very similar sounding and thus often erroneously used interchangeably, they are quite different in meaning. The important thing to remember is that all three terms refer to the same object which gets a different name depending on its position and state (refer Fig. 4). Firstly, a meteoroid travels in space while both meteor and meteorite will be found within

[84] NASA Space Place, "What is a Comet?", https://spaceplace.nasa.gov/comets/en.

[85] NASA Solar System Exploration, "Comets", https://solarsystem.nasa.gov/asteroids-comets-and-meteors/comets/overview/?page=0&per_page=40&order=name+asc&search=&condition_1=102%3Aparent_id&condition_2=comet%3Abody_type%3Ailike.

[86] NASA Space Place, "What is a Comet?", https://spaceplace.nasa.gov/comets/en.

[87] NASA, "Hubble Confirms Largest Comet Nucleus Ever Seen", 12 April 2022, https://www.nasa.gov/feature/goddard/2022/hubble-confirms-largest-comet-nucleus-ever-seen.

[88] Kat Eschner, "Scientists Didn't Believe in Meteorites Until 1803", 26 April 2017, https://www.smithsonianmag.com/smart-news/1803-rain-rocks-helped-establish-existence-meteorites-180963017.

[89] Meteorite Times, "A July 1790 Witnessed Fall: Barbotan, France A Holy Grail Met With Disbelief ", https://www.meteorite-times.com/Back_Links/2009/july/Accretion_Desk.htm.

[90] Christies, "Barbotan Meteorite Shower Of 1790—Partial slice of meteorite shower that reportedly killed a herdsman", 2021, https://onlineonly.christies.com/s/deep-impact-martian-lunar-other-rare-meteorites/barbotan-meteorite-shower-1790-partial-slice-meteorite-shower-19/112845.

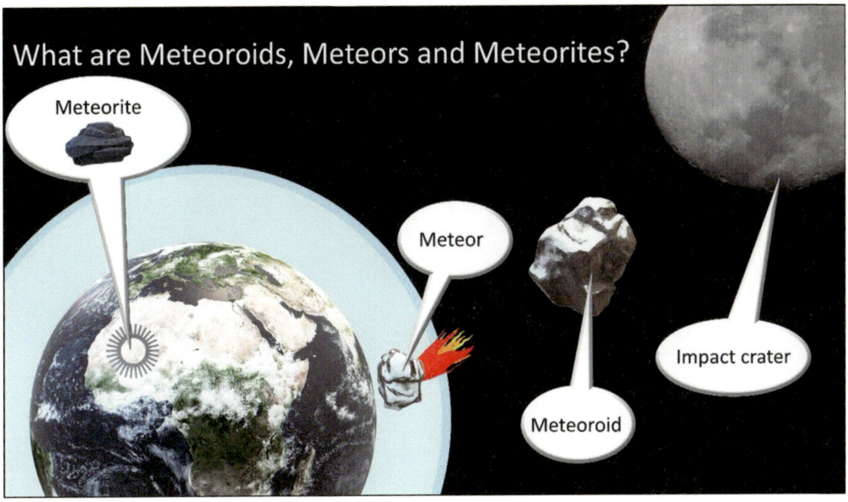

Fig. 4 What are meteoroids, meteors and meteorites? (Graphic courtesy of the author)

the atmosphere of the Earth. Secondly, it can be considered that both meteor and meteorite will represent a morphed state away from the original meteoroid which initially reached the outer atmosphere of Earth.

7 Meteoroid

Very simply put, a meteoroid is a natural object travelling in space (not human made space junk) which can be found in orbit around another object such as Earth, not dissimilar to an asteroid, but much smaller. Whereas asteroids are typically large objects, a meteoroid is much smaller, ranging in size from a speck of dust to that of a smallish asteroid.[91] Meteoroids are typically the by-product of a high-speed impact between travelling space objects (in most cases believed to be between asteroids) but can also involve comets, moons and planets themselves.[92] In Fig. 5 three such scenarios are demonstrated.

A. In this scenario an asteroid collides with a comet travelling through space due to their orbits intersecting at the wrong moment in time. A hypervelocity event

[91] Solar System Exploration, "What's the difference between a meteor, meteoroid, and meteorite?", https://solarsystem.nasa.gov/asteroids-comets-and-meteors/meteors-and-meteorites/overview/?page=0&per_page=40&order=id+asc&search=&condition_1=meteor_shower%3Abody_type.

[92] David J. Eicher, "Where Do Meteorites Come From?", 1 July 2019, https://astronomy.com/magazine/greatest-mysteries/2019/07/18-where-do-meteorites-come-from.

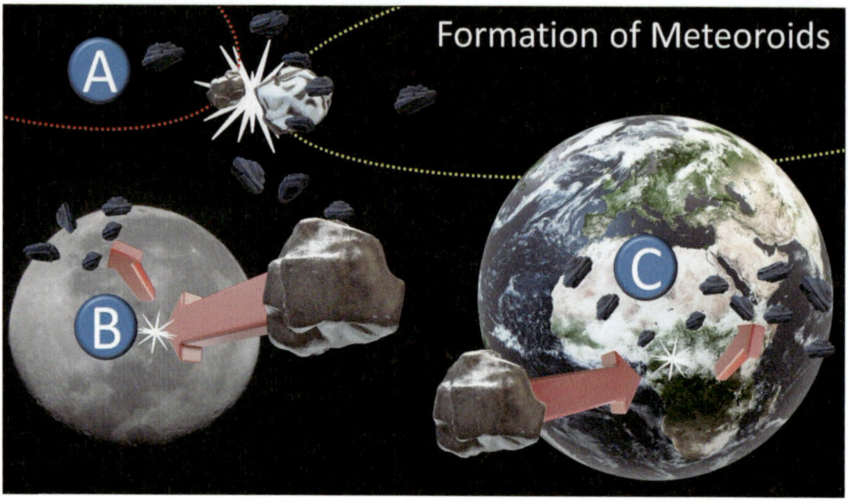

Fig. 5 Meteoroid formation (Graphic courtesy of the author)

such as this can result in thousands of meteoroids being scattered in different directions in space.

B. In this scenario an asteroid collides with an object with a weaker gravitational pull like a planetary moon (e.g., the atmosphere-less Earth Moon). Chunks of debris will be able to escape the impact site easier and deflected into space as meteoroids. In 1979 a small 52,4 g meteorite was found in the Yamato mountains in Japan, known as *Yamato 791,197,* it was the first meteorite proven to be a piece of the moon, subsequently more than fifty similar pieces have been discovered worldwide.[93]
C. Similar than scenario B but in this case the object, an asteroid, collides with a large object with strong gravitational force and an atmosphere such as the planet Earth. In this case it will be more difficult for the pieces of debris deflected from the impact site into space having to contend with both an atmosphere and the gravitational pull. Pieces of Mars have also made its way to Earth, with most of the 175 Mars meteorites recovered thus far, discovered in the North African region.[94]

[93] K. Righter, "Yamato 791,197", *Lunar Meteorite Compendium* 2010, https://curator.jsc.nasa.gov/antmet/pdffiles/f01_yamato-791197v3.pdf.
[94] Jet Propulsion Laboratory,"Mars Meteorites", https://www2.jpl.nasa.gov/snc.

8 Meteor

A meteor is essentially the name given to a meteoroid when it enters the atmosphere of a planet. As this entry takes place at an extremely high speed, friction between the object and the atmospheric components result in the object heating up rapidly—subsequently starting to 'burn-up' in the sky.[95] To an observer viewing this spectacle from the planet surface it will typically present themselves as fireballs darting across the sky with a trail of intense light. Colloquially described as a 'Shooting Star' these objects have probably been seen marvelled and wondered about by humans from the first time intelligent eyes glazed up at the night sky from Earth. Culturally the sight of a 'shooting star' was seen as a very significant event in ancient Greece and Rome it was believed the appearance of the phenomenon was a prediction of either the birth or death or a famous person, and thus could bring with it political uncertainty.[96] On a more practical side though, through the ages, farmers and sailors have also used the direction of 'shooting stars' as tools to aid weather prediction, something that has ironically been take up by modern scientist who are also studying the appearance of these objects to aid understating of high altitude weather behaviour.[97,98]

9 Meteorite

A meteorite is any portion of the original meteoroid, which survives the high-speed fiery trip as meteor all the way to a final impact on the surface of the planet Earth where it typically will leave an 'impact crater' depending on the size of the meteorite. Impact craters are the fingerprints of meteor strikes, forming on the surface of a space object that is hit by another, such as planet Earth being hit by a meteorite—the result is a characteristic circular cavity created by debris of the impact event.[99] There are a number of very good examples on Earth—one of the most well-known example perhaps being 'Barringer Crater', in Arizona State in the USA, also erroneously called 'Meteor Crater' the term meteorite would have been

[95] Solar System Exploration, "What's the difference between a meteor, meteoroid, and meteorite?", https://solarsystem.nasa.gov/asteroids-comets-and-meteors/meteors-and-meteorites/overview/?page=0&per_page=40&order=id+asc&search=&condition_1=meteor_shower%3Abody_type.

[96] Aliya Zuberi, "What shooting stars mean in different cultures", 18 February 2022, https://cutacut.com/2022/02/18/what-shooting-star-mean-in-different-cultures.

[97] Bambi Turner, "10 Superstitions About Stars", 16 April 2021, https://science.howstuffworks.com/10-superstitions-about-stars.htm.

[98] Liv Ragnhild Sjursen, 28 October 2014, "Studying shooting stars helps improve weather forecasting", https://norwegianscitechnews.com/2014/10/studying-shooting-stars-helps-improve-weather-forecasting.

[99] Lunar and Planetary Institute, "Shaping the Planets: Impact Cratering", https://www.lpi.usra.edu/education/explore/shaping_the_planets/impact-cratering.

more appropriate.[100] It is considered the best preserved impact crater on Earth believed to have been formed due to the impact of a 500.000 metric ton object travelling at around 15 km/s, leaving the 1,2 km diameter, 170 m deep indentation featuring 45 m high rim walls, it is arguably the quintessential impact crater.[101] In Norway an impact crater actually became a town. The small town of Nördlingen, was built within the *Ries* crater, with the perimeter wall of the town built on the inner ring of the crater which is now obviously not visible anymore.[102]

It is estimates that just over 6.000 meteorites land on Earth's surface annually of which only a very small percentage are recovered. A total of approximately 60.000 or so meteorites have been recovered on Earth by the time of the writing of this work.[103,104] It is believed that most meteorites recovered from the surface of Earth are the products of asteroid collisions, though examples believed to originate from the Moon and Mars have also been recovered. Meteorites are important sources of scientific learning as it can be used as reference to compare its analysis with data obtained from other sources to make new discoveries or confirm speculation. A good example being, meteorites found on Earth could be confirmed as having originated from the Earth moon by scientifically comparing it to the samples brought from the lunar surface by the Apollo[105] and Luna[106] programs.[107]

According to NASA, it is estimated around 100.000 kg of small particles collide with the atmosphere of Earth daily though, typically only objects larger than 25 m will make it to the surface of the Earth. The frequency of civilization threatening impacts are measured in millions of years.[108] It is believed it was an object colliding with the Earth which led to a devastating extinction event taking out an estimated 75% of life on Earth at the time, including the dinosaur species. Believed to have been more than 11 km wide it impacted Earth in what is now the Yucatán

[100] American Museum of Natural History, "Barringer Crater", https://www.amnh.org/exhibitions/permanent/meteorites/meteorite-impacts/meteor-crater/barringer-crater.

[101] David J. Eicher, "Take a trip down to the bottom of the best-preserved impact crater on Earth", 2017, https://astronomy.com/bonus/crater.

[102] Daisy Dobrijevic, "10 Earth impact craters you must see", 29 October 2021, https://www.space.com/10-earth-impact-craters-you-should-visit

[103] Richard A. Lovett, "Earth hit by 17 meteors a day", 2 May 2019, https://cosmosmagazine.com/space/earth-hit-by-17-meteors-a-day.

[104] Jet Propulsion Laboratory,"Mars Meteorites", https://www2.jpl.nasa.gov/snc.

[105] NASA, "The Apollo Missions", https://www.nasa.gov/mission_pages/apollo/missions/index.html.

[106] Anatoly Zak, "50 Years Later, the Soviet Union's Luna Program Might Get a Reboot", 18 July 2021, https://www.popularmechanics.com/space/moon-mars/a36984208/soviet-luna-program-history.

[107] Rhian H. Jones, "IX.B Lunar Meteorites", 2003, Encyclopedia of Physical Science and Technology. 3rd edition Edited by Robert A. Meyers (Ramtech Limited: Tarzana, 2003).

[108] NASA, "Asteroid Fast Facts", 31 March 2014, https://www.nasa.gov/mission_pages/asteroids/overview/fastfacts.html.

peninsula and leaving as a reminder a more than 200 km wide impact crater named Chicxulub.[109]

The size and velocity of the objects that created the giant impact craters on Earth, such as Barringer, would almost certainly have resulted in dislodged pieces of planet Earth being ejected at high speed into space. Researchers using simulation models estimated the dinosaur destroying Chicxulub event could have propelled pieces of Earth so far as to place it in a collision course with the moon Europa orbiting the gas giant Jupiter.[110] Organic life, or at least the type we are familiar with on Earth, needs three broad basic components namely—the right building-blocks, some form of energy source and liquid water.[111] Evidence points toward the existence of an abundance of liquid water under the ice surface of the moon Europa (believed to be double the combined volume of all the oceans on Earth, though it's just a quarter of the size).[112] As such Europa could in theory be a place where life that originated on Earth could find a place to develop within the vast ocean, provided energy and essential other components of life are there. Should an Earth object from the Chicxulub event thus have made it all the way to Europa, it could have penetrated its icy surface and ended up in the liquid water trapped below. In 2024, a NASA mission will be launched to Jupiter, on arrival the orbiter *Europa Clipper* will commence an orbit around the Jovian planet enabling it to make regular close flybys of the moon Europa collecting data to indicate the possibility of life.[113] Another mission will be sent to Saturn's moon Titan, as the second largest natural satellite in the solar system, it is also the only one known with a dense atmosphere and surface bodies of water similar to Earth such as lakes and oceans.[114] A rotorcraft named the *Titan Dragonfly*, expected to arrive on the moon Titan no earlier than the mid 2030's, will explore the celestial body searching for the presence of the tell-tale chemical building blocks required for life.[115]

But what if it is possible to create a human driven version of panspermia where a contingent of humans leave the Earth with the express purpose of seeding the universe with human life? In the following section a possible scenario will be explored. Taking the concept of panspermia as a template, where the process is

[109] The University of Texas at Austin, "Asteroid Dust Found in Crater Closes Case of Dinosaur Extinction", https://news.utexas.edu/2021/02/24/asteroid-dust-found-in-crater-closes-case-of-dinosaur-extinction.

[110] Rachel J.Worth, Steinn Sigurdsson, Christopher H. House, "Seeding life on the moons of the outer planets via lithopanspermia", Astrobiology 13.12 (2013): 1155–1165.

[111] Christopher P. McKay, "Requirements and limits for life in the context of exoplanets", Proceedings of the National Academy of Sciences 111.35 (2014): 12,628–12,633.

[112] NASA, "Europa, ocean moon", https://solarsystem.nasa.gov/moons/jupiter-moons/europa/in-depth.

[113] NASA, "Europa Clipper", https://www.jpl.nasa.gov/missions/europa-clipper.

[114] NASA, "Titan", https://solarsystem.nasa.gov/moons/saturn-moons/titan/in-depth.

[115] NASA, "NASA's Dragonfly Will Fly Around Titan Looking for Origins, Signs of Life", 27 June 2019, https://www.nasa.gov/press-release/nasas-dragonfly-will-fly-around-titan-looking-for-origins-signs-of-life.

seen as one where several random events need to be in place for the process to be a success, one without a guaranteed outcome but a 'work in progress' built on the hope that the correct circumstances conducive to settle human life will be found somewhere in the universe at some faraway point in the future.

9.1 Directed Panspermia

Directed panspermia, as the name implies, is a version of panspermia where the life-seeding organisms are deliberately directed to a planet such as Earth—it is something that is planned, not a random event.[116] The idea of this section is to look at this concept of directed panspermia from the perspective of humanity becoming a truly explorative spacefaring species. The following subsection introduces a conceptual presentation of just how such a human version of directed panspermia could play out. It explores the idea of an intergenerational human space trek using natural space objects, akin to a cosmic hitchhike, using a strategy borrowed from the lifecycle of a virus.

9.1.1 Human Cosmic Hitchhikers

Hitchhiking is a practice by which a travelling party obtains a travel opportunity from passing vehicles typically without payment; mostly associated with road travel. The 'hitchhiker' will use a hand sign indicating the desire for a ride. In Douglas Adams' Sci-Fi Classic, *The Hitchhiker's Guide to the Galaxy*, the main character, Arthur Dent, unwittingly befriended an alien named Ford Prefect who was on a research assignment to Earth.[117] On being notified of the Earth's imminent destruction to make way for a new 'Hyperspace Bypass', the alien Ford Prefect along with the protagonist narrowly escapes the destruction of Earth by hitching a ride on a passing space freighter. Using an 'Electronic Thumb' a device which allows the owner to 'hitch a ride' on any spaceship passing within reach of these devices which also somehow manages the transport of the travelling party to the vessel used in the onward journey.

For the concept which is presented in this work, a form of hitchhiking will also be involved, though the spacefaring humans will be using a combination of means, to execute what will be a multi-generational interstellar seeding journey. In a nutshell, the basic proposal is simplistic—a group of space farers leave the Earth in a traditional spacecraft, they find a suitable object, such as an asteroid, to hitch a ride with. The travellers embed in the asteroid (which it consumes subsequently for survival resources and to build new craft to accommodate a growing future population who will then repeat the process). In fact, the process is very similar to the lifecycle of the virus discussed in Sect. 3 illustrated in Table 5 for reference.

[116] Francis H.C. Crick, Leslie E. Orgel, "Directed panspermia", Icarus 19.3 (1973): 341–346.
[117] Douglas Adams, *The Hitchhiker's Guide to the Galaxy*, (New York: Harmony Books, 1980).

Table 5 Viral strategy of the Human Cosmic Hitchhiker concept

Virus	Human cosmic hitchhikers
Infects suitable host organism	Spacecraft finds suitable asteroid
Make copies of itself using host resource	Embedded in asteroid building new craft
Copies of the original virus and mutations now leaves host	Different spacecraft evolve, leaves the asteroid
Process repeats itself	Process repeats itself

Incidentally, the notion of hitching a ride on a celestial object as means of travel is also not a new one. In Native American folklore there is the legend of a Hopi mother and child using a ride on a 'meteoroid' to visit the sun in order to answer a question.[118]

The following three sections will attempt to describe the initial three phases of the which will be referred to simply as Human Cosmic Hitchhiking (HCH) in the context of this work.

Phase One

The initiation phase, a human crew leaves Earth using a spacecraft which will be used to locate a suitable asteroid to travel with. In Fig. 6 a graphic representation is provided to illustrate what is broadly expected to be a four phased process. This phase thus represents the equivalent of the Donor Object Escape (DOE) in the ballistic panspermia theory.

A. An initial human crew departs the Earth in a spacecraft powered by conventional chemical rocket technology. At the time of this launch a suitable asteroid which will be used as cosmic ride would have already been identified
B. Their initial destination is a rendezvous point on the Earth's moon.
C. On arrival on at the transit station on the Earth moon, the crew will transfer into another spacecraft for the onward journey to the asteroid. The journey to the asteroid will commence using a hybrid propulsion system—a conventional rocket type system for launch and landing operations and a nuclear version for the space trek.
D. The spacecraft will adjust its own orbit en-route to the target object as required (indicated by the blue dotted line). On arrival at the orbit intersection point with the asteroid, the crew will land their spacecraft on a predetermined location on the surface of the asteroid, which will conclude the first phase.

[118] ICT Staff, "Orionid Meteor Shower and the Great Leader Tecumseh ", 13 September 2018, https://indiancountrytoday.com/archive/orionid-meteor-shower-and-the-great-leader-tecumseh.

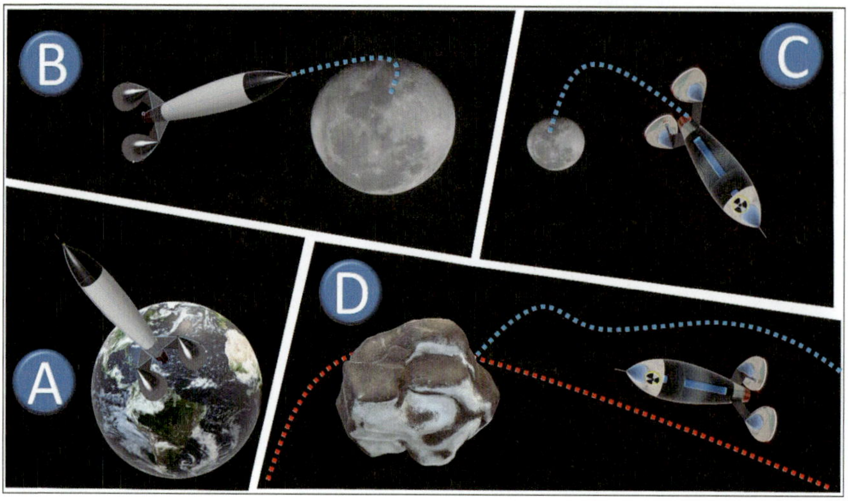

Fig. 6 Human cosmic hitchhiking phase one (Graphic courtesy of the author)

Phase Two

During this phase, the spacecraft lands on the selected asteroid and starts to embed itself into the asteroid. The basic idea is that the asteroid will protect the spacecraft and its human crew from the hostile space environment, similar to a lithopanspermic object. The craft will be equipped with all the necessary equipment to bore into the asteroid. Asteroids will be selected based on the availability of resources which can be used to construct new spacecraft as well as to supply the necessary survival resources whilst the asteroid and its human crew travels through space looking for a planet 'seeding' opportunity. In terms of the ballistic panspermia theory this phase thus represents the equivalent of the Space Wandering Phase (SWP). Figure 7 provides a graphic representation of the expected process.

A. Spacecraft lands on the asteroid.
B. Using specialised equipment, it starts to bore into the asteroid
C. For the duration of the space trek whilst looking for a seeding opportunity, the asteroid is mined for resources creating more spacecraft for the human population which now will be growing.

Phase Three

Phase three represents the phase where there is a transition, when members of the travelling HCH leave the original object to find another opportunity, with the goal of ultimately finding a planet on which human life could thrive (as depicted in Fig. 8). If all goes according to plan, the initial crew will be able to mine the asteroid successfully and use its resources to maintain their internal habitat whilst building new spaceships for what should be an expanding population. As such, the asteroid represents a finite resource which at some future point will

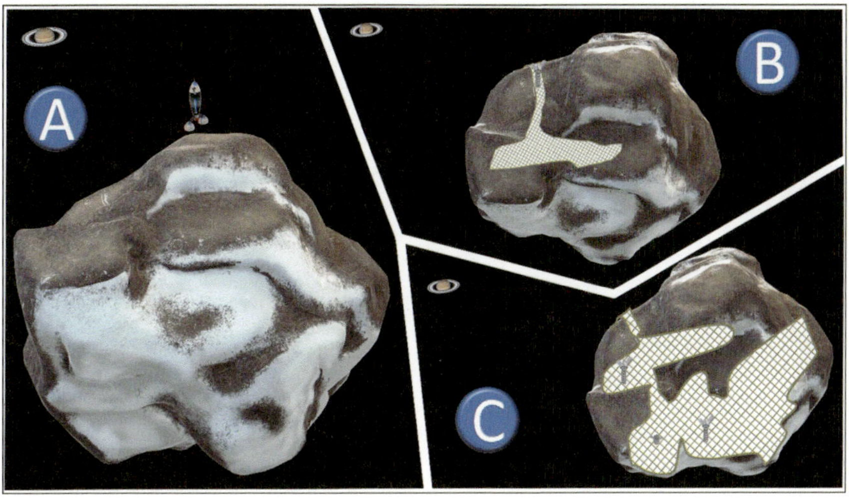

Fig. 7 Human cosmic hitchhiking phase two (Graphic courtesy of the author)

become useless to the travellers. The search for new opportunity thus needs to be a constant driving force. Opportunity could present itself in two ways, firstly another space object presents itself as a travel opportunity or a suitable planet to seed with human life is found. In the first instance a portion of the inhabitants can now leave the original asteroid in a spacecraft that was constructed for such a purpose and embed in the new object. The remainder of the inhabitants continuing their initial journey using the original object. The more times the original crew would be able to 'shed' some of its population growth when a suitable object presents itself the better the chances of successfully spreading human life into the universe. By 'shedding' part of its population every so often, the original asteroid will remain useful for its inhabitants much longer and the population demand on available resource will be more sustainable. This process, like the viral cycle where mutations do happen to aid future survival, almost certainly will lead to changes and different outcomes in development of tools and survival techniques for the different generations 'shedding' from the original. The mission could be seen as complete when the HCH reach the vicinity of a host planet suitable for human life to thrive, and which can be reached with the spacecraft available to the asteroid dwellers.

A. New spacecraft technology has developed from the available resources which will be used to transfer portions of the populations to new opportunities.
B. A portion of the population transfers to a suitable asteroid to commence a separate journey.
C. Population reached a planet suitable for habitation and transfers all or just some of the population to the planet using a capable spacecraft.

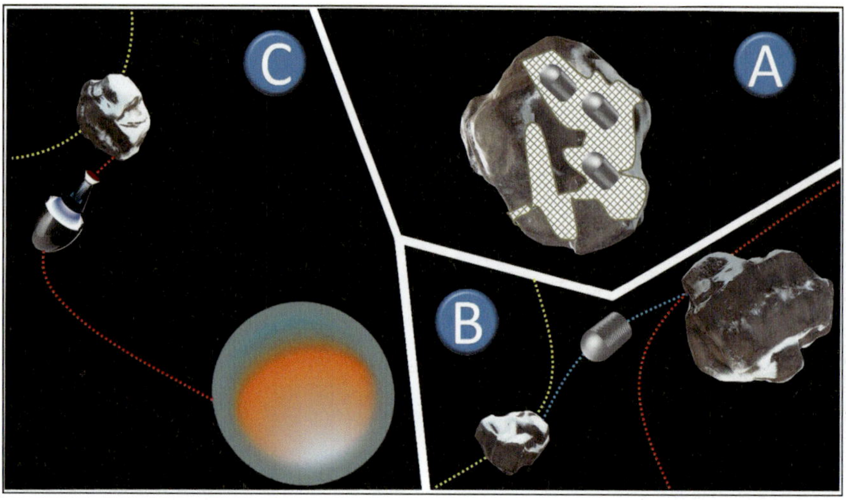

Fig. 8 Human cosmic hitchhiking phase three (Graphic courtesy of the author)

This final phase can thus be seen as analogous with a completed viral infectious cycle by which the virus infected the host, replicated itself successfully and 'shed' into the environment ready for another.

When a new crew embeds into a new asteroid it is quite possible that this asteroid might have unique circumstances forcing the crew to produce a completely different looking class of craft than the one they have arrived in. This could be in the way of novel minerals leading to new construction techniques resulting in a more exotic looking craft. To come back the virus analogy, the design of the spacecraft will 'mutate' if there is an opportunity for improvement and with the new 'host' environment providing new means. It is expected that with each subsequent generation of crew finding their own HCH object, that later the spacecraft that will be produced will change as well, as depicted in Fig. 9. In the graphic five hypothetical shapes of spacecraft are depicted which have evolved over multiple generations of travellers using the HCH way of space exploration.

A. The original first-generation craft in this case depicted as looking not dissimilar to a rocket but in this case the craft contains nuclear propulsion in addition to housing traditional chemical rocket type propulsion.
B. This second-generation spacecraft still resembles the rocket design whence it came from, but now using a much more advanced ionic thrust propulsion system compliments of the availability of novel elements.
C. A more exotic looking next-generation spacecraft, where the availability of exotic minerals made possible the construction of a mesh which can be used to harness cosmic energy and convert it into propulsion for the journey to the destination.

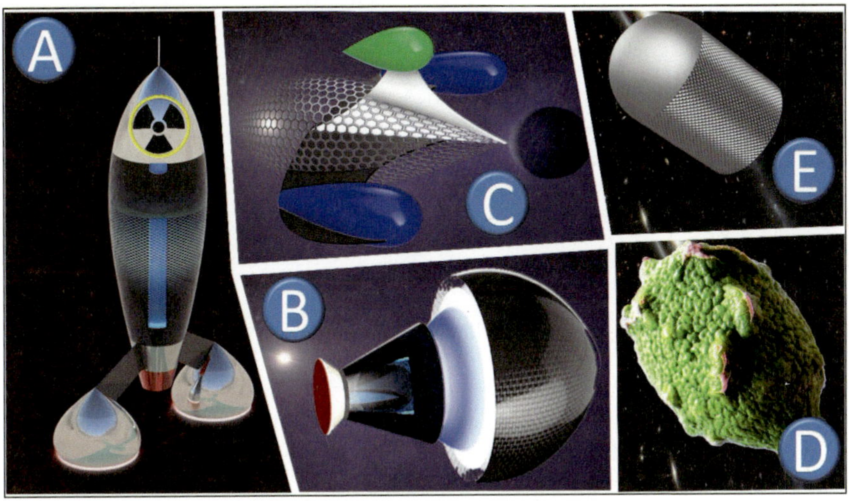

Fig. 9 Human Cosmic Hitchhiking Spacecraft evolution. (Graphic courtesy of the author)

D. At some point into the far-far future one of the nth-generation of spacecraft might be a living entity, protecting its human crew in a self-healing, self-propelling organism capable of obtaining its energy from space. Capable of converting 'stellar-dust' into energy and resources akin to photosynthesis of plants on Earth. This type of craft might become a destination in itself for its crew as it will be able to travel space in perpetuity, able of spawning copies of itself it will thus become a HCH in its own right.
E. At some point one of the crews might design a very simplistic looking craft, but one that will be capable of traversing vast distances of intergalactic space almost instantaneously, using dematerialisation and rematerialisation technology at which point the process of human lithopanspermia will become, in theory, instantaneous.

10 Conclusion

> We choose to go to the moon in this decade and do the other things, not because they are easy, but because they are hard—John Fitzgerald Kennedy.

On 12 September 1962 USA President J.F. Kennedy uttered those famous words at Rice University in Houston Texas, in what became known as the 'We Choose the Moon' speech, to rally and inspire a nation to achieve, what was at that time, an impossible dream. Only five years before on 4 October 1957 the space age officially began when a small artificial satellite called Sputnik 1 was successfully launched into low earth orbit by the then Union of Soviet Socialist Republics (USSR).[119]

Space travel is indeed hard, and there have been many failures and sad loss of life in the shaping of the technology which ultimately would translate into many great achievements. It is hard to think that it took just over five decades since the launch of Sputnik 1 for Voyager 1 to become the first interstellar spacecraft. Whereas Sputnik 1 returned to Earth to burn up in the atmosphere after a mere three months, both Voyagers 1 & 2 were still operational in 2022, nearly 45 years after its launch and a good deal longer than their planned 5-year mission window.[120,121] Voyager 1 became the first craft launched from Earth to exit our solar system in August of 2012[122] while it took its twin Voyager 2 a couple of years longer only exiting on 5 November 2018.[123] The Voyagers were also meant to be ambassadors, mounted on the bus of each Voyager spacecraft is a 'Golden Record' which contains several images, a selection of music and sounds of nature with greetings in various languages including some extinct ones.[124] Should the spacecraft be intercepted by an interstellar spacefaring race, they will be able to retrieve the analogue encoded information by following some basic instructions engraved on the protective gold-plated aluminum cover of the record using the supplied analogue cartridge and needle. Each record is incidentally also covered with a thin coating of Uranium-238 which one could consider as its 'ship clock' counting the days elapsed since it was made.[125]

And just say after travelling thousands of years on their lonely journey, the Voyagers never manage to bump into an alien race to share the 'Golden Records', but rather one day it just makes a soft crashlanding on some exotic faraway planet. It so happens that this planet has the ideal conditions to host biological life, now there always remains the possibility that somehow a scrap of biological life such as a tardigrade or lichen did manage to hitchhike along for the Voyager's epic ride. Assuming it survived the journey and the crashlanding, this life-seeding organism now starts to evolve into what will eventually become a curious race of intelligent beings. Then one day in the far future these intelligent decedents of the original

---

[119] NASA, "Sputnik and the Dawn of the Space Age", https://history.nasa.gov/sputnik.html.

[120] Ethan Siegel, "This Is Why Sputnik Crashed Back To Earth After Only 3 Months ", https://www.forbes.com/sites/startswithabang/2018/11/15/this-is-why-sputnik-crashed-back-to-earth-after-only-3-months/?sh=24ad4ad26dc1.

[121] Megan Gannon," 35-Year-Old Voyager 2 Probe Is NASA's Longest Mission Ever ", 21 August 2012, https://www.space.com/17201-voyager-2-nasa-longest-mission.html.

[122] Solar System Exploration, "Voyager 1", 4 February 2012, https://solarsystem.nasa.gov/missions/voyager-1/in-depth.

[123] Michelle Starr, "t's Official: Voyager 2 Has Left Our Solar System And Is Sailing in Interstellar Space ", 4 November 2019, https://www.sciencealert.com/voyager-2-is-officially-out-of-the-solar-system-and-sailing-through-interstellar-space.

[124] Jet Propulsion Laboratory, "What are the contents of the Golden Record?", https://voyager.jpl.nasa.gov/golden-record/whats-on-the-record.

[125] Jet Propulsion Laboratory, "The Golden Record Cover", https://voyager.jpl.nasa.gov/golden-record/golden-record-cover.

hitchhiking lifeform now a happens upon the "Golden Record", its Uranium-238 coating with its half-life of 4,5 billion years slowly ticking away ready to pose intriguing questions to its discoverers.

Humanity might still be very far away from embarking on an epic space life-seeding journey as presented in this work, but that is not to say it will not happen somewhere in the perhaps not too distant future. Every great journey somewhere starts with an idea after all.

Christoffel Kotze established a boutique technology advisory company in 2012 after a successful corporate career spanning two decades. This company specialises in providing assistance to digital transformation projects within organizations, with a special interest in the use of technology resources to support sustainable development. Current research interests include space technology, dematerialisation through digital transformation and exploring solutions to the 'digital divide.' Qualifications include MPhil (Space Science) University of Cape Town, Bachelor of Commerce Honours (Information Systems)—University of Cape Town, Bachelor of Science (Physiology & Microbiology)—University of Pretoria, Diploma in Data-Metrics (Computer Science) University of South Africa, a number of strategies focussed executive management courses at the Graduate School of Business from the University of Cape Town. ISACA Certified in the Governance of Enterprise IT (CGEIT), TOGAF 9 Certified (Enterprise Architecture). He has published several publications most of which with a focus on the practical application of space technology, with a special interest in its application with regards to progressing the UN Sustainable Development Goals.

Africa: Home of Space Art and Indigenous Astronomy

Barbara Amelia King

Abstract

Africa: Home of Space Art and Indigenous Astronomy examines the symbiotic relationship between art and science, referencing Africa's legacy as the home of space art and indigenous astronomy in relation to its future as an emerging space faring continent. The immense relief felt by Africans as the last of the countries dismantled colonial domination throughout the past sixty years, awakened a long-suppressed sense of optimistic creativity in the generations that followed. This surge of enthusiasm and hope for their future is embodied in the moniker 'Afrofuturism.' This inquiry peers back into the genesis of Space Art and Indigenous Astronomy, and looks forward into an Afrocentric future as constructed using the science fiction genre. It examines the influence that African experiences have had on the world view of contemporary African visual artists, writers and filmmakers. Within the last several decades space art and space science have become two of society's most influential disciplines, thus the value of art relevant to space and society has developed accordingly. This article interprets the dynamics of Afrofuturism through the arts as Africa enters Space 2.0.

1 Afrofuturism

The successive dismantling of colonial domination by African countries, one by one, within the last sixty years, has awakened a long dormant sense of individual creativity in Generation X, the Born Frees, the Millennials, and those coming after. One dramatic way this surge of unleashed innovation has manifested itself through

B. A. King (✉)
Peachtree Parkway Arts, Peachtree City, USA
e-mail: Barbara.king.z@gmail.com

A. Froehlich (ed.), *Outer Space and Popular Culture*, Southern Space Studies,
https://doi.org/10.1007/978-3-031-25340-9_4

the arts is by the creation of Afrocentric science fiction underpinned by the concept of Afrofuturism. African artists respond similarly to science fiction like the rest of the modern world, albeit through an African sensibility that actively links societal inclusion and environmental concerns in their world view.

This article examines Afrofuturism and its rise to global popularity through science fiction in visual arts, literature and film. Afrofuturism is a philosophy of empowerment based on equitable principles adjusted for a modern age scenario. Similar to science fiction, Afrofuturism seeks the high ground, by saying: let's dream big, let's do the impossible, lets level the playing fields, let's get super-powers, lets vanquish the enemy, equalize cultures, and let class, gender, and race prejudice disappear along with tribalism and all the other 'isms' in society. Artwork, from its observational position, has generally employed the Humanities as an ethical guideline for its social commentary, especially pertinent for new space civilizations which will be built from scratch. The Sci Fi genre is also a platform from which to draw together disparate factions of society, using creativity and innovation to their individual and mutual advantages.

Oscar Wilde's 1889 axiom that 'Life Imitates Art, far more than Art Imitates Life,' can be applied to the phenomena that many space exploration concepts first exhibited by artists actually have been manifested into astronautical realities. The interplay between space science and space art has become a self-perpetuating phenomenon as more innovation morphs into reality; and space technology, in turn, creates new possibilities for artistic expression for artists, scientists and the general public. This article clarifies why that is so by examining the purpose of space art to amaze, inspire and instruct and also details the function of artists as interpreters of science, as social commentators, as agents to bridge cultural divides and as space researchers.[1]

The author seeks to construct the relationship of space art to space science and society by focusing on examples of African visual art, graphic art, science fiction novels and films. This research also reinforces the notion that space art raises public awareness of space-oriented innovation and applications to stimulate sociopolitical change that improves living standards. Finally, this enquiry explores the legacy of Africa's space art and indigenous astronomy, contrasted with Africa's future as a powerhouse entering the newest epoch of Space exploration.

2 Romantics, Dreamers, Explorers

The expanse of the Universe, those celestial bodies it contains and the space in between them, is the basis of space art and also the nexus between space art and space science. As demonstrated by the earliest evidence on cave walls depicting celestial events, to art that has been placed on the Moon, Mars or in orbit in current

[1] King, BA, *Space Art & Space Science,* Master of Philosophy thesis, University of Cape Town, South Africa, 2020.

day, artists strive to formulate that which they can see in space via technology, and that which they can only imagine in lieu of any representation. 'Space Art' describes a relatively new genre, which derives its name and classification from its subject matter. The genre gained its provenance from depicting cosmology fables, science fiction concepts, philosophies, the scope of the humanities, and each idea developed with space science itself. *The International Association of Astronomical Artists*, perhaps the only, and most certainly, one of the most prestigious guilds of its kind, supply a recent definition of the subsections of space art as applied since the advent of space flight.

> Space Art or Astronomical Art, is the genre of modern artistic expression emerging from knowledge and ideas associated with outer space, both as a source of inspiration and as a means for visualizing and promoting space travel. Like other genres of artistic creation, Space Art has many facets and encompasses realism, impressionism, hardware, sculpture, abstract imagery, and even zoological art.[2]

Artists in all genres continue benefitting from what space technology has been offering: more precise, smaller, robust, shiny, fun and less expensive equipment, materials, and data for artistic interpretation. The resolution of photographic images has improved by quantum leaps, providing never before seen locations from which to draw inspiration. There has been increased access to space information and space itself as artists are invited to participate more closely in many more scientific endeavors, all of which culminate in an increased impetus to produce art for space. Following on, the space science community, astronauts, scientists, engineers and influencers of every ilk, are expanding their ability in creative arenas by deploying art modalities to further develop their innovative thinking and polymathic skills. Dr Carolyn Porco, planetary scientist and imaging team leader for Cassini, is aware of the synergistic relationship between space scientists and space artists:

> Space Art...a noble genre of art reminds us of the delight and inspiration we have been given by astronomical artists all over the world and throughout time. It is a marvelous celebration of their soaring imaginations, technical skills and artistic talents. And for me, art will hereafter be a means to recall the special bond that joins them to us scientists. For, like scientists, astronomical artist are romantics, dreamers and explorers, ever yearning, ever seeking, ever hopeful.[3]

Even more so in this prolonged information age, what the public understands about space and space technology is increasingly conveyed using creative arts platforms produced by Porco's artistic romantics, dreamers and explorers. The idea of space is no longer merely relegated to imaginative entertainment, it is now potentially

[2] International Assn. of Astronomical Artists, home page, https://iaaa.org/what-is-space-art/ (accessed 14 July 2022).

[3] 2 Ron Miller, *The Art of Space: The History of Space Art, from the Earliest Visions to the Graphics of the Modern Era*, (Zenith Press 2014, ISBN 978-0760346563) 6.

life-saving, and therefore; personal, political, and commercial. Most often the public, educational, governmental and scientific communities are informed of new art or science via the media in an ever-increasing variety of platforms. Encouraged by civil society entrepreneurs, the agenda of the New Space movement is concentrated on societal inclusion and environmental concerns as anchors for space research and sustainability. The social agenda regarding space and humanities' relationship to it began many millennia ago in Africa.

But what those artists would have seen then is much different than what space artists see today. Nearly two million years ago when Homo Erectus evolved, (Fig. 1) the Milky Way would have been leaner, drawing into itself all the dust, asteroids, planets, gases, stars and smaller galaxies that it could entice. It would have shown more of its red-hot interior but not the black hole at its heart. Artists then, would have seen more planets and multitudes of stars that appeared overly large and luminous in the sky before they, too, were absorbed to eventually become the Milky Way with which we are familiar. (Fig. 2).

Fig. 1 Barbara King, *The Tremendous African Sky.* Digital Painting, 2022 Courtesy of the artist

Fig. 2 Barbara King, *What They Might Have Seen.* Digital Painting, 2022 Courtesy of the artist

2.1 Space Art and Indigenous Astronomy

Each generation of African artists over time have watched the Universe unfold and were undoubtedly influenced by its impact on themselves and their civilization. Let us envision ourselves as the artist and try to interpret what the universe may have looked and felt like. The enormous skies, without the impediment of buildings, must have been awe-inspiring, and yet the formidable vastness must have also been simultaneously paralyzing. Yet, night after night, year after year, generation after generation for thousands of years on end, each person likely had to develop and maintain their sense of being and balance when rendered absolutely physically powerless, and psychologically insignificant by nature (Fig. 1). To maintain that balance, Africans made a continued study of the sky, noting the calendar of days, and the astronomic events that occurred therein (Fig. 2). To an artist, a question would be just how extraordinarily dramatic the sky would have appeared one hundred thousand years or more ago, before electricity obscured the view. How impactful, awesome, emotional, absolutely glorious and vividly ominous every day and evening might have been as the sky moved through the night and turned into day. Yet instinct was stronger than fear, so artists observed, drew, painted and etched their way through the seemingly endless morass of the unknown, and demonstrated their psychological state of mind by inventing space art derived from a cosmological narrative. The African cosmological narrative; then, can be considered the first science fiction story which spawned the genre many thousands of years old and still going strong.

2.1.1 The Cradle of Space Art

South Africa is referred to as the Cradle of Humankind, the home of the first human beings, and by extension, the first space artists and the first indigenous, cultural astronomers. Evidence suggests that our earliest ancestors deeply valued the power of drawing, painting and engraving and admired the artist's ability to replicate what they were seeing, thinking and feeling.

Archeological evidence unearthed in 2011, along with subsequent studies of the find in 2018, reveal that artists have been producing artworks since the dawn of human history. One hundred thousand years BCE (as we know now, but this could be even earlier as future digs may reveal) artists left behind their abalone shell paint-grinding equipment and compound mixtures of pigment in the Blombos Cave near Cape Town, South Africa. Along with it, hidden under layers of sand dune, were ochre shards etched with abstract designs, bone paintbrushes and quartzite grinding stones.[4] Archaeologists dated the abalone shell containers and their pigments as 100,000 years of age. Another significance of this find is that it is the first known instance of a deliberate mix of various mineral oxides with binders and curative liquids, and as such, dates artists as the earliest users of elementary chemistry.[5]

The opinion of visual artist and MIT engineering polymath Robert E. Mueller is that artists discovered spirits in the stars and invented 'gods' to help control their physical and metaphysical worlds. Further, that recording celestial bodies and events in paint were completed in the hopes of affecting change, foretelling manifesting the future and bringing some semblance of control over their destiny. Art was the early human's total science. It was their logic system. They concretized supernatural powers through an artistic attitude, to cope with the realities which they simultaneously embodied and explained.[6]

The earliest known engraving, a zig-zag pattern incised on a fresh water shell

[4] Stephanie Pappas, "Oldest Human Paint-Making Studio Discovered in Cave", *Live Science,* https://www.livescience.com/16538-oldest-human-paint-studio.html (accessed 26 March 2019).

[5] Ian Sample, "*Neanderthals—not modern humans—were first artists on Earth, experts claim*", *The Guardian,* www.theguardian.com/science/2018/feb/22/neanderthals-not-humans- were-first-artists-on-earth-experts-claim (accessed 28 March 2019).

[6] Robert E. Mueller, *The Science of Art: The Cybernetics of Creative Communications,* (Oxford University Press, 1967, ISBN 9,780,853,911,081) 22.

from Trinil, Java, was found in layers dated to 540 000 years ago. In terms of drawings, recent findings determined that painted representations in three caves of the Iberian Peninsula were 64,000 years old—this would mean they were produced by Neanderthals. Thus, the drawing on the Blombos silcrete flake is the oldest drawing by Homo Sapiens ever found."[7]

A further find by the same team at the same site in 2018 yielded what is called the first abstract design/ drawing by Homo Sapiens, Modern Humans, dated at approximately 63,000 years ago. It is a series of cross hatched marks that are repeated in ochre crayon in several places in the cave, and found in several different archeological layers.[8] One can imagine the impact of these current revelations on the space artists or indigenous astronomer today, knowing that what they are instinctively compelled to do in their respective fields has been an innate tendency in human kind from the beginning. Additionally, that the art and science they often have suffered to do, has the legitimacy of existing longer than any other known discipline of human kind.

That their art practice alone has carried on the ideals, and activities of humankind over a hundred thousand years is a revelation not to be taken lightly. The drive of artists to practice art was born of human beings' innate desire to express what they saw, heard, felt, thought and wanted known. Among the images are scenes recording cultural events, natural sciences, fantastical beings and the celestial phenomenon of the skies. Art in the form of drawing, painting and engraving is the earliest attempt at a physical rendering of the relationship between humans and their abstract, uncommunicable universe.

2.1.2 Ancient Astronomy, Indigenous Cosmology

Astronomer, astrophysicist and cosmologist Professor Thebe Medupe, host of the *Cosmic Africa* documentary series, concurs that Africa's astronomers have an ancient history, deeply rooted in mathematical and cultural astronomy which illuminated the continent's ancient story. From the myths and legends attached to San folklore; and the calendar system of the Dogan of Mali based on the phases of the moon to the mini megalithic Stonehenge of Nabta Playa in the Sahara, and the manuscripts at Timbuktu, Madupe is convinced that mathematical astronomy existed in Africa centuries before modern scientific institutions were established,

---

[7] Henshilwood, C, van Niekerk, Karen Loise, *South Africa's Blombos cave is home to the earliest drawing by a human, The Conversation* 12 September 2018, https://www.theconversation.com/profiles/christopher-henshilwood-22291.

[8] Henshilwood, C. et al., *An abstract drawing from the 73,000-year-old levels at Blombos Cave, South Africa,* Nature, 12 September 2018.

and also agrees that African artwork demonstrates how humans integrated their knowledge of art and science.[9]

The Smithsonian Museum of African Art in Washington DC addressed cosmological issues when curating the 2012 *African Cosmos: Stellar Arts* exhibition to demonstrate that astronomy is culturally embedded by virtue of artists throughout time via engaging in dialogue about the relevance of the sky and celestial bodies in their artistic practice.[10] This exhibition exemplified the connections between art and science, and how they each explored the cosmos, and influenced African art and ritual.[11]

The artwork of South African Gavin Jantjes was featured in the *African Cosmos: Stellar Arts* exhibition and adorns the cover of the exhibition's book. Using an analogy of African cosmology taken from the word 'zulu,' which translates to 'the space above your head, the heavens,' he created the *Zulu* series, based on ancient cosmology of San/Khoi San rock art, mythology and astronomy. Reminiscent of the style of the earliest San art rock paintings, Jantjes referenced a San myth about the formation of the cosmos, which recounts how a young girl dancing around an evening fire threw glowing embers into the night sky and formed the Milky Way and the stars.[12] Jantjes' 'Untitled' can be seen at https://africa.si.edu/exhibits/gavinjantjes.html. Author and artist Barbara King creates yet another rendition of the same cosmological story two decades after Jantjes, and many years after the San. The telling of the story never gets old. (Fig. 3).

There are always new tools by which an artist can express the emotion and movement of a scene. In *The Dancer,* (Fig. 3) three modern digital painting and drawing programs were combined to invent the scenario of a San dancer forming the Milky Way from ash, and the stars from embers of her fire. She was positioned as having fun and feeling free as she flings the embers into space. It is significant that this African cosmology was not born from war, dominance, monsters, death or desperation as it is in other cultures. It was born from freedom of movement, from the creative instincts of a young girl playing on the beach who has the remarkable foresight to invent the stars, orchestrate light in the sky, and begin eternity.[13]

Fast forward to the twenty-first century, wherein the creative arts and space science and developed their revolutionary thinking and disruptive behavior alongside the progression of technology. The features of space art to amaze, inspire

---

[9] Allison Keyes, "Millennia of Stargazing At African Cosmos," *National Public Radio,* 28 October 2012, https://www.npr.org/2012/10/28/157736107/millennia-of-stargazing-at-african-cosmos-exhibit (accessed 16 March 2020).

[10] Christine Mullen Creamer, "Curating African Cosmos/Stellar Arts" *Anthropology News,* 11 July 2019, https://www.anthropology-mews.org/index.php/2019/07/11/curating-african-socmos-stellar-arts (accessed 12 February 2020).

[11] Swingler, H., *Africa's ancient astronomers made their mark in culture, folklore, University of Cape Town News* www.news.uct.ac.za/ 02 December 2014.

[12] *African Cosmos: Stellar Arts,* https://africa.si.edu/exhibits/cosmos/intro, 2014 (accessed 4 March 2020).

[13] 'The Girl Who Made the Stars', Specimens of Bushman Folklore 1911 Anabezi Camp, Zimbabwe Wildlife Trust, https://anabezi.com/the-girl-who-made-stars/ (accessed 20 July 2022).

Fig. 3 BA King, '*The Dancer.*' Digital Painting, 2020, Courtesy of the artist

and instruct has always been essential, but is now examined more intensely when development of new civilizations on multi planets is underway. Space art has been the fulcrum from which many philosophers, artists and scientists have forged their ideas of universality.

3 Capturing the Universe

Art about space has had a profound effect on many scientists, engineers, writers, filmmakers, photographers and artists who have encountered it. Space Art's collective inquiry into the Universe and what might reside there is a subject that Professor Roger Malina, physicist, astronomer and executive editor of international

Art/Sci/Tech journal *Leonardo*, also credits for ushering in the space age. The creative collective used their agency to push their vision of space forward. It is a quote often used, and always worth repetition.

> The space age was possible because for centuries the cultural imagination was fed by artists, writers and musicians who dreamed of human activities in space. Now with the end of the cold war the role of artists and writers is again crucial in defining our future vision of space and will once again be instrumental in incorporating the facts and discoveries of the space age into the cultural imagination.[14]

Similar in approach to Roger Malina, Ghanaian science fiction author Jonathan Dotse recognizes a symbiotic relationship between technological innovation and science fiction in the Western world, wherein he surmises that increased technological development in Africa will provoke a corresponding surge of science fiction. The surge will do more than just entertain, in that science fiction prompts a discourse that is speculative and takes a long view of societal development.[15]

> I had lived in Ghana my whole life. It had been changing my whole life, but I didn't really feel the magnitude of change until I left and returned; and I was really shocked by how much had changed in three years. The arts scene was way more vibrant than when I'd left. There was a lot more optimism, a lot more creativity. And that gave me the basis for being able to imagine the future.[16]

Science fiction novels, films and visual art serve as a vchicle to raise awareness of indigenous and modern technological innovation, and the hybridization of each. In turn, sci-fi art inspires sociopolitical changes and improved living standards because the story lines and narratives speculate about a future society benefitting from space-based applications. Those narratives include the latest philosophy and characteristic societal customs and conventions revolving around humanistic viewpoints.

4 Interpreting Science and Humanities

An artistic eye must be employed to contextualize the ever-increasing mountains of camera images and sensing data from space when aiming them for scientific and non-scientific audiences. Artists take over where the camera leaves off to graphically comprehend and elucidate data which often times has never been encountered

---

[14] Stephen Wilson, *Information Arts: Intersections of Art, Society and Technology*, (MIT University Press, 1990, ISBN 9,780,262,731,584), 261.

[15] Sherelle Jacobs, "The weird worlds of African Sci Fi", *African Business Magazine*, 20 June 2015, https://www:africanbusinessmagazine.com/uncategorised/continental/the-weird-worlds-of-african-scifi (accessed 5 February 2020).

[16] Geoff Ryman, *Jonathan Dotse, 100 African Writers*, Strange Horizons, 11 July 2022: https://www.strangehorizons.com/non-fiction/ (accessed 14 July 2022).

before. Their task is to translate numbers, equations, dots, bars, audio, zig zags and words into images that illustrate stories, and accurately represent the measurements data indicates.[17]

These are scientific functions in which artists excel. From an artist's point of view, though, that is only one portion of space art's domain. Astronomical artists also deduce plausible scenarios from the available facts when they interpret scientific data in the creation of art.[18] In 1990, astronomical artist and planetary scientist William K. Hartmann theorized that artists acted as interpreters of science, as social commentators, as agents bridging cultural and political divides, and as researchers in the disciplines of space art and space science.[19]

Artists have readily taken to the task, as evidenced by the prodigious amount of imagery illustrating comic books, magazines, science fiction books, art work and film posters that have been created throughout the last four hundred years. All of these renderings required imaginative illustration, thereby providing artists with a platform for their work, and an opportunity for successive generations of readers and movie goers to be engaged with the concept of space exploration.[20] Those generations are the ones who now sit at the engineers' desks and are in the science labs, operate ground control, launch the hardware, manipulate the software, fund and design science missions, live on the ISS, and are preparing to do the same on the Gateway and beyond.

By 2017 the functions of space art widened even further in service to space science as observed by Gisela Williams, in 'Are Artists the New Interpreters of Scientific Innovation?' The resurgence of interest in inviting artists to observe, learn and work with mainstream government agencies, institutions, entrepreneurs, scientists is consistently on the rise. Now that it has been recognized by the space industry that creativity underpins technological innovation, and innovation has become science's lifeblood, collaborations and residencies are becoming popularized. "Every company seems to be launching an experimental lab that is meant to foster innovation through the cross-fertilization of ideas in a variety of disciplines, including the creative arts."[21] Gerfried Stocker, the artistic director of Ars Electronica in Austria, is adamant that the role of artists has become that of cultural

[17] BA King, *The Limitless Horizons of Space Art,* Outer Space & Popular Culture, Influences and Interrelations, Part 2, Southern Space Studies, Springer 2022, pp. 53–74, https://doi.org/10.1007/978-3-030-91786-9.

[18] Andrew Dickson, "What Artists Would Do if They Could Fly to the Moon," *New York Times,* https://www.nytimes.com/2018/09/24/arts/design/artists-moon-spacex.html (accessed 12 October 2019).

[19] William Hartmann, Andrei Sokolov, Ron Miller, Vitaly Myagkov, *In the Stream of Stars: The Soviet/American Space Art Book,* (Workman Publishing, 1990, ISBN 9780894808753), 132–142.

[20] BA King, *The Limitless Horizons of Space Art,* Outer Space & Popular Culture, Influences and Interrelations, Part 2, Southern Space Studies, Springer 2022, pp. 53–74. https://doi.org/10.1007/978-3-030-91786-9.

[21] Gisela Williams, *Are Artists the New Interpreters of Scientific Innovation*? New York Times, Sept 12, 2017 www.nytimes.com/2017/09/12/t-magazine/art/artist-residency-science.html (accessed 11 July 2022).

missionaries as well as the overseers of ethical and moral developments.[22] Mike Stubbs, the director of Liverpool's Foundation for Art and Creative Technology agrees:

> Science is too important to leave to the scientists. Science has kind of become a new church, but it's clear now that technology has not been applied to everyone in society to their benefit. We need voices from the arts and sociocultural disciplines to provoke important debates.[23]

And Stubbs is quite right to be wary of an all-encompassing juggernaut when cultures are vulnerable to being hijacked, and governments subject to the obsessions of a ruling party. Those observations, coupled with the fact that media content shapes public perception globally to a greater degree each day, does indeed transform media artists into essential societal players as interpreters of scientific information. Film, television, radio, blogs and streaming have the shock and awe factor that instantaneously sways the public narrative about the value of space to society. The Art/Science relationship has been duly exploited by the film industry, who use space technology to dramatize their cinematographic offerings in entertainment and education.

The remainder of this article is dedicated to giving an African context to Hartmann's functions of art, to Williams' multi-disciplinary creative labs, to Stocker's concerns regarding ethics oversight, and Stubbs' interest in applying a socio/cultural balance regarding science. Artists in Post-Colonial Africa leverage their ancient past with their new science fiction future.

4.1 Entering Africa

The vast African continent with so much unseen 'otherworldly' territory is a feast for the imagination. It's cultures, color, landscapes and wildlife has intrigued authors, visual artists and filmmakers for centuries, and it is the perfect slate upon which to depict science fiction scenarios. That the science fiction genre is once removed from real life non-fiction drama renders its format as an objective stage upon which to personify abstract human qualities and enshrine cultural hopes and dreams. The following are examples of films, novels, and artwork with Afrocentric content that envision life on Earth and in space.

4.1.1 Interpreting Alien Species: District 9

District 9 is a science fiction alien invasion movie shot on location in Soweto, South Africa in 2009. It speaks to a universal inquiry; how will Earthlings relate to alien species when and if they meet? Can they get along and resist the tendency toward discrimination, domination, racism, gender bias, xenophobia, and all the

[22] Ibid.
[23] Ibid.

other phobias? These elements are incorporated into the *District 9* story line as Aliens crash-land their space ship in a 1980's Soweto, a township originally engineered to segregate black from white. In *District 9* the victims of discrimination are insectile, prawn-like and are immediately segregated from society, reminiscent of the racial segregation that was rampant in South Africa until the early 1990's.[24] Ironically, the film's message was that because humans behaved so badly the aliens were struggling to escape from Earth, not to overtake it.

Johannesburg born first time director and co-writer Neill Blomkamp employed African ingenuity to produce a blockbuster film on an extraordinarily small budget, in a short production window, with an unknown cast, and in a dangerous location. The blend of computer-generated imagery and visual effects, contrasted with the rawness of the location, garnered excellent technical and content reviews, and the film shot to the top of the US box office in its opening weekend. District 9 was nominated for Oscars under the categories of best picture, film editing, visual effects and screenplay and captured the Best International Film award from the Academy of Science Fiction, Fantasy and Horror Films, along with a dozen more accolades from within the industry.[25] Its awards demonstrated that an African space context situated in an African landscape within the science fiction genre merits more than just a topical interest globally. Its success engendered an upsurge of creative interest about the continent, and set the stage for other African science fiction story lines to come, such as *Black Panther,* produced nearly a decade later.

4.1.2 Interpreting Africa's Sci-Fi Potential: Black Panther

In 2018 Africa became a pivotal part of the science fiction universe of Marvel Comics with its release of the African-centric *Black Panther* movie. Its fictional story line revolves around African characters living in a mythical African Kingdom of Wakanda. Though only a few images were actually filmed in Africa compared to its locations in Argentina, South Korea, and Georgia, USA, it was the magnificent topography of Zambia, Uganda and South Africa that legitimized the authenticity of the film's landscapes. The might and majesty of both the real and imaginary African landscape and cultural life helped elevate the Sci Fi category and garnered Marvel their first ever Oscars for best production, best costume and original score.[26]

The authentic scenery, the use of both African and African-American actors, and an African centric science fiction script catapulted *Black Panther* into becoming the highest grossing Super Hero movie of all time in the USA. *Black Panther's*

---

[24] CNN, "Sci-Fi Apartheid Film District 9 Opens in South Africa", *CNN Entertainment,* 31 August 2009, https://www.cnn.com/2009/SHOWBIZ/Movies/08/31/district.nine.blomkamp/index.html (accessed 25 February 2020).

[25] Internet Movie Data Base, "District 9 Awards", *IMDb* (2019) https://www.imdb.com/title/tt1136608/awards (accessed 5 March 2020).

[26] Paul Meeken. "How Much of Black Panther Was Shot in Africa? Not A Ton", *Heavy.com Entertainment*, 18 February 2018, https://heavy.com/entertainment/2018/02/black-panther-shot-in-africa-on-location (accessed 15 March 2022).

overwhelming triumph indicates that African cultures have heroic content that easily encompass ancient to modern timeframes, and also span the divide between science and science fiction.

4.1.3 Interpreting Afrofuturism: Binti

First generation Nigerian-American Nnedi Okorafor has turned to African folklore and fantasy as a basis for her short stories and novels, which she has subsequently translated into the genres of films and comic books. A prolific Afrofuturist science fiction author, she is the winner of the Nebula, World Fantasy and Macmillan Writers of Africa awards, as well as the recipient of the Wole Soyinka prize for African literature, and is an Arthur C. Clarke and British Science Fiction Association finalist. Okorafor has also written *Black Panther* scripts for Marvel Comics, and is currently adapting her novels featuring the film's female special forces warriors, the Dora Milaje, into a movie script.

Okorafor's Afrofuturistic trilogy *Binti* is set in a technologically and socially advanced future, with the drama revolving around multi-cultural astronauts living off their home worlds, attempting to integrate their cultural legacies into a platform from which to carve out their future in space. The trilogy is scheduled to be transformed into a script for a TV series produced by Hulu, co-written by Okorafor.[27]

This particular sampling of African stories is not extensive, and represents just the beginning. More are under way for Afrofuturists, as Okorafor has successfully negotiated with Home Box Office (HBO) to adapt her 2010 post-apocalyptic novel *Who Fears Death.* It is a setting which serves as a potent call to action regarding the societal issues of civil strife including genocide, weaponized rape, and other circumstances that plague women specifically, in this instance, in a war-torn African country. Additionally, art imitates life as Okorafor's own memoir *Broken Places and Outer Spaces* is also currently being developed into a film.

The storyline in *District 9* satirizes the cultures that embody xenophobia, *Black Panther*, gives heroes and sheroes nearly mythical prowess, *Wakanda Forever* and *Binti* take on loyalty, strength, sense of self and self-preservation, African style. Big, bold, leading the pack. In the end the people's will is not broken, they win against the odds. Africa is not a victim.

To be in a land that has always been, like Africa; that has always been peopled by your people, that has thrown off the yoke of oppression to embrace the concept that all is possible is a dream, it's a surreal, condition wherein anything can actually happen. Sci Fi is a brilliant platform to reinvent an entire culture. African sensibilities of time, space, imagery, culture and art are reinterpreted by artist and audience alike. Afrofuturism is nourished, native truths are told, native stories retold, hidden perspectives arise from whichever character is relating the

[27] Carmen Guiba, "Queen of Sci-fi Author Nnedi Okorafor Book 'Binti' Is Finally Being Filmed by Hulu", *Africa Futura Newsletter*, www.africafutura.com/queen-of-sci-fi-author-nnedi-okorafor-book-binti-is-finally-being-filmed-by-hulu/ 17 January 2020 (accessed 15 March 2022).

story. The story's climax will be what Africans want and need, and developing an end game strategy will be the triumph of each artist, writer and filmmaker across genres. Underpinned by the victory and resolve manifested in the Afrofuturistic attitude, eyes are also now turned towards nurturing the Humanities, and evaluating the prosperity promised by the benefits of Space.

4.2 Encompassing the Humanities

The act of expressing culture on the walls, then appreciating the results, gave rise to the artist being regarded as a social commentator in modern times. If archeologists are stunned by the breadth, depth and color of the arts they see now after thousands of years of environmental degradation, imagine what the art sites would have seemed like to the originators...absolutely divine. Art did not stop at handprints, stick people and animal look-alikes, artists kept developing more complex scenes and imagery. Many scientists and anthropologists fail to comprehend what an emotional medium art is for the artists and the viewer. Art has the potential of bringing every human emotion and cultural concept to the fore. It is in the purview of historians, scientists and artists to interpret how space art, indigenous astronomy, and cosmology, has influenced the Humanities as culture and civilization evolved. Africa's own traditional humanistic world view, Ubuntu, is the basis of Afrofuturist concepts of social equality and cultural inclusivity.

Science fiction is a platform upon which to play out those possibilities. And it is working. The global spotlight on Space 2.0 has popularized the art of science fiction for a new generation of artists to create their Afrofuturist realities. Three artists who 'Africanized' space by their specific inquiries concerning humanitarianism, governance, globalization pursuant to Africa's future in space, are Jacque Njeri, Jean-Bosco 'Shula' Monsengo, and Yinka Shonibare.

4.2.1 Jacque Njeri

African photographers, graphic artists, illustrators and digital painters are experimenting with science fiction, virtual reality, and digital photo editing to address and redress historical narratives, comment about current social issues, and imagine the future potentials. Kikuyu art director, and digital painter Jacque Njeri positions the Maasai tribe of Kenya and Tanzania as space explorers. Her *MaaSci* series of digital collages (a combination of Maasai and Sci-Fi) depict traditional explorations (Fig. 4), and societal issues such as female empowerment and cultural diversity, projected in a science fiction future. (Fig. 5).

> Afrofuturism is a portrayal of our rich cultural aesthetic through tech, science and fantasy themes projected as extraterrestrial realities. It is also a statement that the inclusion of black

Fig. 4 Jacque Njeri, 'The Elders', MaaSci digital collage. Maasai warriors exploring the interstellar void aboard a flying saucer, Digital Illustration, 2017. Courtesy of the Artist

> characters in Sci-Fi is dependent on authors, artists, filmmakers and other content creators immersing themselves into the field and making that a possibility.[28]

In Njeri's work one can assume the Maasai as symbolic of many African cultures, involving themselves in space on their own terms, without precluding their cultural interest or identifying factors. The elders in Fig. 4 would need to acquire a roomier, oxygenated, protective covering capable of diverting atmospheric radiation if they were travelling in real time, and not in an imaginary tableau. But that does not stop the artist's imaginative inquiry from making a vivid point that traditional semi-nomadic Maasai see themselves as travelers who are interested in and comfortable in space. The vibrant color and style of neck jewelry worn by the 'Shestory' astronaut not only represents her culture and gender, but also distinguishes a female sensibility and cultural flair when contrasted the Star Wars inspired grey mono-color planet behind her. (Fig. 5). The image is also jolting in contrast to sixty years of photographs in which white men from a super power country are shown as the only conquerors of space. Njeri's artwork is a reminder

[28] Lindsay Samson, *Jacque Njeri's vibrant graphic art reimagines the Maasai people in outer space,* Design Indaba 27 March 2018, https://www.designindaba.com/search?keywords=jacque+njeri.

Fig. 5 Jacque Njeri. 'Shestory', MaaSci, Female Maasai Afronaut on a planetary body, Digital Collage 2017. Courtesy of the Artist

that very, very soon, astronauts from different genders and cultures from developed and emerging space nations expect to be exploring space and living on celestial bodies.

4.2.2 Jean-Bosco 'Shula' Monsengo

The narratives of painter Jean-Bosco 'Shula' Monsengo, of the Democratic Republic of the Congo, reflect the importance of education, and public awareness about critical social events, environmental issues, good governance and international politics. The surrealist aspects of his Afronaut paintings, (Fig. 6) and his bright colours transcend the traditional way of depicting the world, yet embody both modernity and tradition by pairing twenty-first century satellites with African prints and traditional sculptures. (Fig. 7) Shula's Afro-futuristic work addresses his optimism in Africa's capacity to carefully manage the influx of globalization and Africa's ability to chart its own course through space and technology as it emerges on the world stage.[29]

[29] "Shula Monsengo", *African Art Gallery Hong Kong*, 2019, https://www.africart.com.hk/monsengo-shula (accessed 20 February 2020).

Fig. 6 Shula Monsengo. '*Roi Satellite*, 2014, Acrylique sur toile,140 × 140 cm. Image: Galerie Magnin-A, Paris. Here artists are pictured on an extra vehicular activity painting a satellite. Courtesy of the artist

4.2.3 Yinka Shonibare, CBE

Yinka Shonibare, CBE, RA, is a British-Nigerian artist living in the United Kingdom. His work explores cultural identity, colonialism and post-colonialism within the contemporary context of globalization. A Shonibare work is immediately discernable because of the brightly coloured bespoke fabric he designs, then drapes onto his perfectly sculpted (molded so artfully that one might think they are actually breathing) mannikins.

'Planets in my Head' features the mannikin of a young girl straining to peer into a telescope aimed at the stars as she delicately balances on a wooden box to reach the eyepiece. (Fig. 8) The thrill of discovery and the awe of physicists, astronomers and scholars worldwide who have gone before revolve around her head. Archimedes, Galileo, Einstein, Hawking, Shen Kuo, Ibn Rushd (Averroes), Yoshio Nishina and Brahmagupta are emblazoned on the globe that substitutes for her head. It is at once a serene and intimate portrayal of a young girl's curiosity about space, following in the footsteps of her foremothers, and an iconic image of women's success at entering a male dominated world. Shonibare's work is multi-media and multidisciplinary. His characters are constructed in Western

Fig. 7 Shula Monsengo. '*Ata ndele mokili ekobaluka/Sooner or later, the world will change.*' 2014. Acrylique et paillettes sur toile,130 × 200 cm. Image: Galerie Magnin-A, Paris. Courtesy of the artist

tableaus, that easily lead the viewer to contrast social bias, economics and political relationships between Africa, the US and Europe.

Shonibare's 'Space Walk Astronaut' installation depicts astronauts at the frontier of the American Apollo era in which a female and male astronaut mannikins, dressed in brightly bespoke Dutch/Malaysian/African print space suits are engaged in a space walk outside their spacecraft (Fig. 9). As pioneers, they Africanize Earth's outward push of expansion, exploration, and colonization. And they do it in a fashion forward African style with matching boots, backpacks and tethers. In the Space Walk sculpture, the African identity of their country of origin very visible through their clothing as the astronauts are reliant upon on one another as they are attempting one of the most dangerous activities in space. This sculpture sets up a fundamental law of Space wherein class, race, gender, economic backgrounds are rendered irrelevant. Life in Outer Space is the equalizer of all things cultural and political.

4.3 Encountering Inconvenient Truths

Art has an unparalleled ability to educate the masses and provide a platform from which to coalesce support from many quarters. Mutual appreciation of space and art can prompt disparate entities towards common goals, and assist viewers as they

Fig. 8 Yinka Shonibare. 'Planets in my Head, Physics' Sculpture, 2010. Mannequin, Dutch wax printed cotton leather and fiberglass. 127 × 66.99 × 49.85 cm © Yinka Shonibare CBE. All Rights Reserved, DACS/ARS, 2022

encounter uncomfortable historical narratives that art brings front and center. Art serves as a platform for mediation, as a bridge over troubled cultural or political waters or, conversely, art can stand in protest to the status quo. Following are two examples of how art, artists and their communities intervened in a specific cultural/scientific impasse, with varying success.

To accommodate the vastly improved technology of astronomical devices in modern times, it is imperative to build telescopes on much larger tracks of land, and locate them further from rapidly sprawling, well-lit populated areas. To find this land, searches often extend into environmentally sensitive and/or indigenous lands, or both. All the more problematic a situation in many parts of Africa. South Africa in particular, who's colonial and Apartheid past has festered into a cultural rift that is exacerbated by economic injustice and political unrest. The author would add that it is also land that artists cannot access any further to continue drawing, painting engraving or experiencing, which creates an unnecessary break in the ancient/modern astronomical art continuum.

4.3.1 The Indigenous and Modern Astronomy

One notable example of art attempting to bridge the communication gap between indigenous rights versus the needs of modern astronomy took place in 2014 in

Fig. 9 Yinka Shonibare. 'Space Walk Astronauts, 2002. Screen printed cotton fabric, fibreglass, plywood, vinyl, plastic, steel. Astronauts each: 212 × 63 x 56 cm (83.5 × 22 x 22in) Spaceship ca. 370 length x 153 cm diameter (200 length x 99in diameter) © Yinka Shonibare CBE. All Rights Reserved, DACS/Artimage 2022. Image courtesy Stephen Friedman Gallery, London. (Space ship not shown)

South Africa and Australia. The desired locations for the proposed builds of the Square Kilometer Array (SKA) resulted in a clash between indigenous culture and modern science which demanded accommodation. Artists served as mediators to help bridge the gap. To indigenous artists of Australia's aboriginal Yamaji and South Africa's /Xam-speaking San met with SKA astronomers to find common ground. The exchanges resulted in a meeting of the minds: that the benefits of science must be shared for the benefit of humanity. The results of this encounter were celebrated in the *Shared Skies* art exhibition in 2014, reminding the viewers that space has inspired all people, everywhere, prior to recorded history.[30]

The Australian Yamaji pieces for *Shared Sky* were painted in a traditional Aboriginal dot style, (Fig. 10) and the South African needlework tapestries of the San incorporated visual motifs that stretch back to antiquity as well as reflecting the SKA dishes as sculptures on the landscape.[31] *Shared Sky* opened in Australia in 2014, traveled to South Africa, then toured European venues until 2018, continuing its mission was to bring together many nations around two sites in Australia

[30] Square Kilometer Array, 'Shared Sky: The SKA's indigenous Astronomy/Art exhibition", *SKA Telescope Home Page,* http://www.skatelescope.org/shared -sky (accessed 17 March 2022).
[31] Ibid.

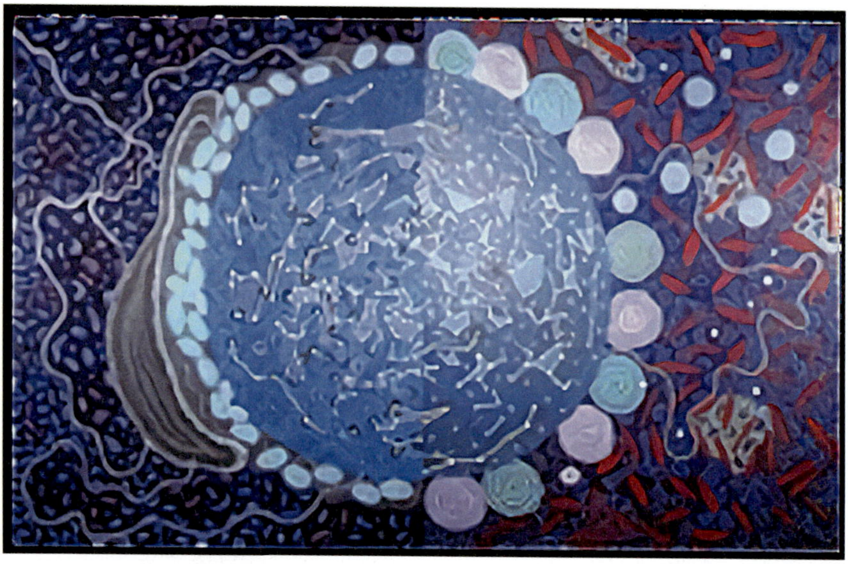

Fig. 10 *Ilgali Inyayimanha* (Shared Sky), painting, artists articulated the theme, which reads in part: "It doesn't matter where we live on Earth, we are all sharing the same sky. Although we may see different things belonging to the sky, we are looking up at the same stars and constellations."[33] 2014. Image: Square Kilometre Array

and South Africa to study the same shared sky developed across countless generations observing the movements of the night.[32] The SKA project went ahead; yet how the State, or the project ultimately addressed the socio/economic inequities for all parts of the indigenous community, and the artists in particular, is a study worth undertaking. The Exhibition has an ongoing presence at https://www.skatelescope.org/shared-sky/.

4.3.2 The Indigenous and the New Kids on the Block

Disenfranchisement of the original inhabitants from the land itself amid the current and pressing need for poverty alleviation, is an inherited situation for the Sutherland telescopes in South Africa, the New Kids on the Block. Visual artist, art director, and big dreamer South African artist Marcus Neustetter, along with his collaborator Bronwyn Lace, took the opportunity of the global attention given to the Southern African Large Telescope's (SALT) successes to give voice to historical and ongoing problems of the indigenous in the area. Neustetter had been embedded with local artists for various other projects, so together they built the town an indigenous version of a telescope within sight of SALT.

[32] Square Kilometer Array, "Shared Sky: The SKA's indigenous Astronomy/Art exhibition" *Square Kilometre Array* www.skatelescope.org/shared-sky/ (accessed 17 March 2020).
[33] Ibid.

Fig. 11 Marcus Neustetter and Bronwyn Lace, *The Sutherland Dome,* 2013 Sutherland, South Africa. Corbel building structure with steel geodesic dome constructed in collaboration with the Sutherland community. Image Marcus Neustetter. Courtesy of the artist

The Sutherland Dome (Fig. 11). is multifaceted. As a monument it is a celebration of humanity, acknowledging the indigenous who were observing the sky here, and the artists who were imaging it, long, long before the telescopes arrived. Thus, the ancient and modern stand juxtaposed in iconic silhouettes. As a protest piece it is a silent witness to the economic injustices of the past, including land grabs, which continue to carve out inequities in the present. The dome also artfully serves the present, the construction is set on a walking trail so that visitors as well as locals can experience the same intimacy with astronomical forces as did their forebears. (Fig. 12).

5 Modus Operandi: Inquiry and Research

Artists play a direct role in scientific research by designing ways to present information, communicate with other scientists, stimulate scientific visualization, and inspire new ideas. Space artists and space scientists have engaged in creative collaborations for decades. However, as space science has become more complex with longer missions and more to make, measure and monitor; so too, has the nature of Art/Sci's relationship become more multifaceted. To create works of scientific validity, artists must acquaint themselves with the laws of science by way of inquiry and research to integrate space science as part of their work. Three

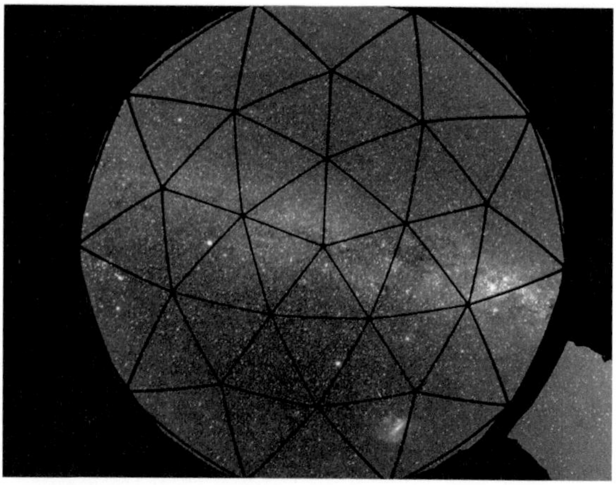

Fig. 12 Inside the Sutherland Dome by Marcus Neustetter and Bronwyn Lace. 2013. Image: Bruno Letarte. Courtesy of the artist

visual artists who root their inquiries and research into space-related technology are South Africans Lyndi Sales, Marcus Neustetter and Barbara Amelia King.

5.1 Inquiry into Logarithms and Sound—Lyndi Sales

Capetonian Lyndi Sales chooses abstraction to portray the Universe and comprehend it mechanics, its chemistry, its light properties and sound waves. Working with hand-woven textiles, sculptural installations and collage materials she dramatically explores the synergy of Art/Science. Sales' artistic inquiry into scientific research begins with rationale and reason, then is abstracted into energy, mass, and the speed of light.

> Looking out into the universe and trying to comprehend what is known as "Dark matter" (which comprises 84% of the universe, yet remains unseen) the abstract seemed appropriate. Looking deep within the microscopic cellular body or out into the galaxy at the platonic solids or atoms of the universe, these structures that make up everything reveal themselves as abstract forms in my work.[34]

Sales' rationale behind the composition of 'Satellite Telescope' (Fig. 13) celebrates the sensational accomplishments of Kenya's Small Astronomical Satellite (SAS - A), the first African Satellite whose mission included a comprehensive survey of the sky for X-ray sources, the discovery/study of a pulsing accretion-powered binary X-Ray source, and the discovery of the first strong candidate for an astrophysical black hole. Nicknamed for the Swahili word for freedom, Uhuru was launched on Kenya's 7th anniversary of Independence, December 12, 1970.

[34] Ashraf Jamal, *In Conversation With Lyndi Sales*, Art Africa, News Articles 2013, 11 September 2013.

Fig. 13 Lyndi Sales, "Satellite Telescope" Multi-media. 2011. Photo credit David Bloomer. Courtesy of the Artist

'Satellite Telescope' is a logarithmic spiral traveling through a space which symbolizes the progressive continuity and rotational creation in macro and microcosms. That continuity and creation represents the repetitive rhythm of life, the cyclical nature of evolution, and being beneath the flux of movement. It also marks the action in a reverse direction of the same power around two poles, and in the two halves of the world. Sales is intrigued by the properties of Earth's magnetic field and its path moving through, in and around Earth and focuses on the notion that a small tremor on Earth has a ripple effect that not only resonates over the planet but also out into the universe as well.[35]

In a particularly fruitful collaboration, Sales visited Texas A&M to embed herself with its team to experience the science of flight and the sensation that sound creates when its barrier is broken. The Sonic Boom Research Team made available its digital research, included airflow experiments, morphing processes, computational fluid dynamics, and renderings. Sales also recorded the sounds of the science labs. Armed with this array of influences and her impressions gained on site, Sales created two relief wall installations, the first of which mimics the movement of the process, the airflow and the acoustics of a plane going through the sound barrier. The second draws upon the aerodynamics of airflow as Sales distorted and modified a subsonic plane blueprint to create a rhythmic composition. Its three sections reveal the impact the plane received as it hit, then broke through, and finally surpassed the sound barrier. (Fig. 14).

Figure 15 represents the fragmentation of a sound wave after the plane had broken the sound barrier. Here Sales tackles the sonic acoustics of a fragmenting

https://www.artafricamagazine.org/in-conversation-with-lyndi-sales, (accessed 18 July 2022).

[35] Lyndi Sales, Artists Statement for Satellite Telescope, email to author, 19 July 2022.

Fig. 14 Lyndi Sales, "Shape shifting to Transcend Limbo". Multimedia. 2016. Image: Lyndi Sales. Courtesy of the Artist

Fig. 15 Lyndi Sales. 'Transcending Realms. Chaos and Flow, Love and Fear.' Multi-media. 2016. Photo credit: Lyndi Sales, Courtesy of the Artist

sound wave overlayed with recordings of adjacent sounds relating to the endeavor, explanations by team members, and preexisting sonic acoustics. Color, the stuff of light, also shifts on the piece during the day and evening. Interest in Art/Sci collaborations have been building steadily since the 1960's, and have become part and parcel of the multi-disciplinary creative experience in many NewSpace organizations. Because an artist does not have to be a space scientist to comprehend astronomical concepts and employ them in their practice. A well thought out collaboration, and the coalescing of the right team, such as Texas A&M with an artist like Sales, has proven to be a very successful modus operandi for the integration of an artistic inquiry into scientific research.

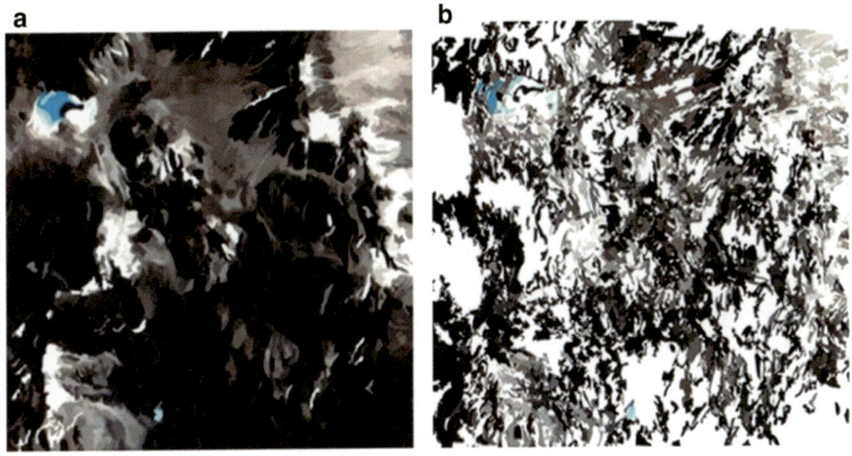

Fig. 16 **a** Marcus Neustetter, *SumbandilaSat,* 2012, digital print. Geographical images based on data received from Sumbandila Sat. Courtesy of the artist. **b** Marcus Neustetter. *Sumbandila Sat—Disrupted Data* 2012, digital print. Geographical images based on data received from Sumbandila Sat. Digital prints, 500mmx500mm.Edition 10. Courtesy of the artist

5.2 Inquiry into Satellite Imagery—Marcus Neustetter

Scientific research has served as the basis of artistic inquiry for community art impresario from Johannesburg, Marcus Neustetter. Neustetter was in the right African country at the right time to become involved with South Africa's first home grown satellite. Having worked closely with NASA since the 1960's to track images in the Southern hemisphere, South Africa is well-versed in space science compared to many other countries on the continent.

In 2009 South Africa's first weather satellite '*SumbandilaSat*' (*Lead the Way* in the Venda language) was launched in Kazakhstan, only to lose contact two years later, a victim of a solar electromagnetic storm. However, the 1,128 high resolution useable images collected while the satellite was still functioning in orbit were impressive, and timely for the use in global disasters in Namibia, Japan and the USA and instrumental in So Africa's park management in general.[36]

To honor SumbandilaSat's accomplishments, Neustetter uses images of its data as a narrative to anthropomorphize the new country's hopes and dreams about using space-based applications to level the playing field in the social/cultural arena. Figures 16a and 17a represent the quality of data received prior to the solar storm. Figures 16b and 17b represent their corollary, the degrading images captured as the satellite tried in vain to continue its mission.

[36] Rebecca Campbell, 'South Africa's Sumbandila satellite has finally fallen back into the atmosphere,' *Engineering News*, Creamer Media, 13 December 2021, www.engineeringnews.co.za/article/south-africas-sumbandila-satellite-has-finally-fallen-back-into-the-atmosphere.

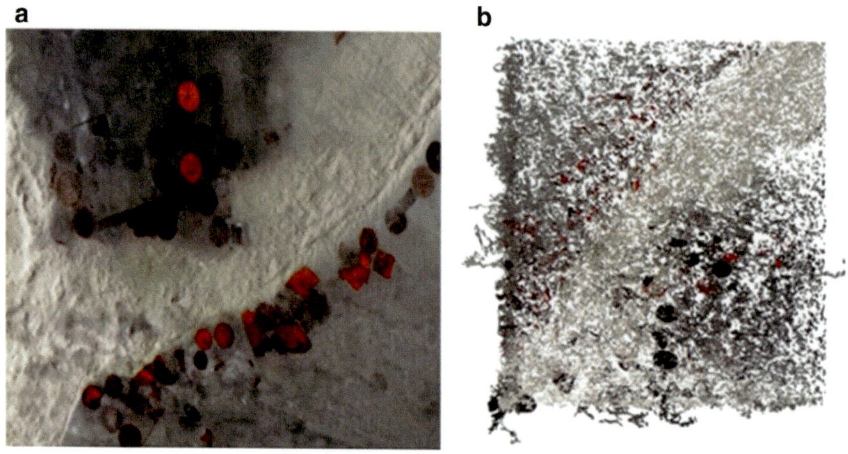

Fig. 17 **a** Marcus Neustetter, *SumbandilaSat,* 2012, digital print. Geographical images based on data received from Sumbandila Sat. Courtesy of the artist. **b** Marcus Neustetter. *Sumbandila Sat—Disrupted Data* 2012, digital print. Geographical images based on data received from Sumbandila Sat. Digital prints, 500mmx500mm.Edition 10. Courtesy of the artist

During its decade of providing some information, but no photographs, Neustetter attempted to contact and follow the defunct space entity, and finally wrote a letter to SumbandilaSat using his artistic interpretation of its own imagery with a morse code message, asking if it might in some way be able to lead yet again. (Fig. 18). But, alas, that was not the satellite's destiny. After following its planned orbit of decay by dint of gravitational pull, Sumbandila Sat was incinerated upon its reentry into Earth's atmosphere on December 10, 2021. Yet, by virtue of the artwork and scientists inspired by it, SumbandilaSat is now one of the African ancestors, its legacy continuing to stimulate curiosity, imagination and new perspectives into space and onto Earth.

5.3 Inquiry into Indigenous Astronomy and Cosmology—Barbara Amelia King

As an artist, author, and space science researcher, three of my digital artworks are included in this article to illustrate indigenous astronomy and cosmology by visualizing space events that would have attracted artists. They are visualized precisely because there are no photographs or illustrations to tell the story of how artists first reacted to the formidable celestial events playing out above their heads. This is the time when artistic expression was born, when the largest, most extraordinary and totally incomprehensible celestial occurrences are raining down, night or day, and they were powerless to understand or control any of it.

Yet, artists took center stage, forming the first images which served as the only way to capture, remember and interpret any logic/reasoning/science of the

Fig. 18 Marcus Neustetter, '*Letter, Lead the Way Again'* 2022. 1 cm x 1 cm x 1 cm. Courtesy of the artist. Microscopic ink drawing on acrylic pages bound in a folded format with laser cut morse code message: *Lead the Way Again* is currently in the Moon Gallery art exhibition aboard the International Space Station

day. The psychological need to make sense of the world, to predict the weather and to calculate astronomical events gave rise to Space Artists and Indigenous Astronomers. The first artist's inquiry into space science began at least 100,000 BCE. I am following suit by instigating artistic inquires with space science as the subject. So are many others. More artists are turning their attention to space art and also to Space Studies degree programs that are being offered at a variety of educational institutions. I followed suit there, too, using my space studies background and digital painting programs to initiate the hybrid Art/Sci concepts behind the Figs. 1, 2, and 3.

'The Tremendous African Sky' (Fig. 1) addresses how the Milky Way came into being, by snacking on its neighbors. 'What They Might Have Seen' (Fig. 2) imagines how big and beautiful items the sky might have seemed one hundred some thousand years ago, how the populace would have been both mesmerized and paralyzed by its terrific power. Interpreting the Universe has intrigued artists and astronomers without fail for these thousands and thousands of years.

'The Dancer' Fig. 3 speaks to devising an African cosmological narrative as a socio/cultural device to explain the Universe and human's relationship to it.

Most, if not all, world cultures seem to have invented what is essentially a speculative science fiction narrative of how Earth, its peoples, plants, animals and the Universe came into being and how they interrelate. Indigenous cosmology reveals the thinking of the day. As time went on many frontiers were crossed, science was born, technology followed, space artists ended up with digital equipment and space-based art applications. Although modern astronomy has a different story of the universe, African cosmologies still inspire the imagination. That imaginative speculation ignited the advent of science fiction, Afrofuturism's platform of choice.

6 Space, the New Frontier for African Arts

Africa was the first frontier for human artists, using the universe as subject matter. Hence, the idea of creating art that goes into space, or creating whist in space is not a far reach, albeit a fabulous new frontier even with space's particular constraints and challenges. Accordingly, welcoming such an endeavor with an African spirit of unity and prosperity for all is logical, certainly now that sharing space data has become a common occurrence among professions and across countries. Africa is home to over a billion people and many of the world's fastest growing economies, so it should come as no surprise that demand for space-based technologies and investment is growing rapidly. Indeed, demand for satellite capacity in sub-Saharan Africa is forecast to double in the next five years.[37] The future looks bright for the African astronautical industry as emerging space agencies, civil space entities, conferences and expositions are sparking with excitement and inventive entrepreneurs are pitching up daily on the continent of Africa.

Several countries have taken a regional lead in space-based data collection, satellite system mission and design, and training through STEAM (Science, Technology, Engineering, Art, Mathematics) programming. Many countries with well-developed space industries are competing to gain African partnerships. The African Union's African Space Policy and Strategy road map was distributed in 2016, and a regional African Space Hub is in the works in South Africa.[38] These developments signal a necessity for artists in the new frontier to ply their trade of image making *for* space, and *in* space. One effort, known as the African Artists Development Program, follows on from other global initiatives of making a cultural statement on space hardware.

6.1 Creating *for* Space

In 2021 Francophone African artists were given an opportunity to compete to have their imagery adorn the fairing of the new generation EUMETSAT (European Organization for the Exploitation of Meteorological Satellites) satellite as it launches into orbit in 2022, thanks to the advocacy of the Paris based African

[37] Jeremy Luedi, 'Africa's space race ready to launch' *Asia by Africa,*18 January 2018, https://www.asiabyafrica.com/point-a-to-a/going-to-space-worth-the-cost (accessed 12 July 2022).

[38] *African Union Heads of State and Government Adopts the African Space Policy and Strategy*, African Union, 31 January 2016, https://au.int/en/pressreleases/20160131-3 (accessed July 2022).

Artists for Development (AAD).[39] In turn, the data from EUMETSAT (which will be stationed over Africa) will be made available to African countries for policy making purposes to assist in reaching the United Nation's Sustainable Development Goals in agricultural development.

Rather than just one artist winning as was originally foreseen, ultimately three finalists were chosen to collaborate on one image to demonstrate the collective power of the continent working together. Michel Ekeba and Geraldine Tobé from the Democratic Republic of the Congo, and Cameroonian Jean-David Nkot were granted an arts residency in Benin and their collaborative result is now ready for transfer onto the nose cone of an Ariane 5 scheduled for launch sometime in 2022.

As the title implies, the Africa of the memory and the Africa of the present are projecting a common future together, and projecting that future into space. Africa is developing into a more fulsome Africa in the space age. The artists demonstrate that Humanism/Ubuntu (prosperity for all) is a collaborative philosophy to take Africa where it needs to be by leveraging the benefits of launches, manufacturing, space-based applications and collaboration with other space faring nations. The three Francophone artists who created the first African art to be blasted into space are now torchbearers and ambassadors for space, to encourage more collaborative space art initiatives within the length and breadth of the Africa. The overarching theme engaging the artists is Space, a new frontier for African creation.

Nkot's contribution evokes an African continent that combines elements referencing Dogon's astronomical symbols, with rivers connecting all countries and roads that translate into lifelines. An imaginary Africa no longer separated by artificial borders. Artist Géraldine Tobé sees the potential of space as a receptacle for dreams, a repository for emotions and a place to design the possibilities of the future. Her figures of African women represent the future of humanity, walking towards the future. Michel Ekéba concentrates on the six climatic zones of the continent, peopled by space walking Afronauts in six colors, to symbolize a United Africa. They dot the canvas to reflect humans in outer space, and signal that every layer of the painting was meant to congeal into one dynamic reflection of the Universe. African Artists for Development harnessed the power of the artistic community to symbolize a commitment of Africa in general, and of African artists in particular, to the development of the continent in line with the United Nations Sustainable Development Goals, 2030. Many more Art/Sci collaborations are expected to follow. 'Memory of Today, Memory of the Future', Jean David Nkot, Géraldine Tobé, Michel Ekéba. NET Collective, Multimedia Painting April 2022 can be viewed at www.africanspaceartproject.com.

[39] African Artists for Development, Press, April 2022, www.africanspaceartproject.com, (accessed May 2022).

6.2 Creating *in* Space

Creating art in the service of space is a growing discipline in emerging and in space empowered countries. However, creating art in space is a quantum leap that is about to occur in a dramatic fashion, with artists having been chosen to be the first private tourists orbiting the moon. Their ability to produce inspirational art during and after the experience will educate billions globally about the intrinsic value of space science. Combining the creative skills of both disciplines, is win–win situation for the arts, for the sciences and ultimately, the global society at large, as this combination can bring forth even more examples of Life imitating Art.

A paradigm shift was set in motion for artists to become mission specialists when the Dear Moon Space Tourist Mission was announced in September of 2018 by Japanese entrepreneur Yusaku Maezawa and SpaceX's Elon Musk. The mission is specifically designed to place artists in a round trip orbit around the moon. Maezawa is inviting eight artists from various creative fields as his mission specialists for the five days, approximately 500,000-mile round trip aboard a SpaceX Starship.

Maezawa conceptualized his universal art project as a philanthropic endeavor to move humanity forward by offering them the intellectual and emotional inspiration of art produced from the trip itself and long afterward.[40] Artists are chosen to be the purveyors of Maezawa's ultimate hope for world peace based on the value of art to space science, and space science to humanity. Should all go as planned, in 2023 artists will occupy a coveted front row seat and take their place in the forefront of the space community by creating art in cislunar space. The African population, and numbers of African space artists are on the rise as it is estimated that by 2050, a quarter of the people on the planet will be African.[41] As Maezawa has conceived this project as an international event, it is not beyond imagination that one or more of the artists chosen to participate might well be from Africa, where viewing, interpreting and interacting with Space all began. A recent update to the contest with over one million entrants, indicates that of the 8 winners, none are from the continent of Africa, and none are traditional space artists in the traditional genres of drawing, painting or sculpture. This selection for such a historic trip further reflects that the astronomical arts and their crucial relationship to space science are not widely understood or acknowledged.

[40] Barbara A. King, *Space Art & Space Science*, Space Studies Mphil dissertation, University of Cape Town, 2019.

[41] Edward Paice, By 2050, a quarter of the world's people will be African-this will shape our future, *The Guardian,* 26 January 2022 07.25 EST, https://www.theguardian.com/global-development/2022/jan/20/, (accessed 20 July 2022).

7 Conclusion

This article demonstrates that African space artists and indigenous astronomers were engaged and active, with the Universe as their subject matter, long before modern science was born. Actually, it has been said that the immensely long continuum of creativity of those space artists and astronomers that gave birth to modern space science. Africa's engagement with the phenomena of space spawned by its first citizens remains unabated despite the prolonged effects by successive waves of tribalism, European aggression, slavery, colonialism and Apartheid.

Indigenous cosmological narratives are the first science fiction stories, the beginning of the genre, and a perfect slate upon which to write sci-fi scenarios based on African perspectives. Africa's own traditional humanistic world view, Ubuntu, is the basis of Afrofuturism, an optimistic philosophy of empowerment based on equitable principles adjusted for the modern age. Afrofuturism as expressed through visual arts, literature and film raise awareness of indigenous and modern technological innovation.

Space artists' role as agents of cultural change, interpreters of science, and as social commentators, is enormously consequential as Afrofuturists link Ubuntu, societal, and environmental concerns with Africa's space agenda. The advent of Space 2.0 has popularized science fiction for a new generation of artists to create their Afrofuturist realities, with eyes turned towards nurturing the Humanities, and stringently evaluating the problems and the prosperity promised by the benefits of Space.

Space was the first frontier for human artists, hence, the idea of creating art that goes into space, or creating whist in space is not a far reach whatsoever. Those dreamers and explorers are now advocating for Art/Sci collaborations through shared inquiries and research, paving the way for artists to create in space. Space artists/scientists can take advantage of their African artistic and astronomical inheritance, where viewing and interpreting space all began. Fitting, then, as the Cradle of Humankind and the Home of Space Art and Indigenous Astronomy, that Africa emerges as a powerhouse in this next space epoch.

Barbara Amelia King is a visual artist in digital and traditional genres. She studied visual arts at the Otis Art Institute followed on with MA in Arts Administration from the Goddard Graduate School, and an MPhil in space studies from the University of Cape Town. She recently came to the US after three decades of producing cultural/historical media in South Africa. A space art researcher and writer, King's art illustrates the science underlying universal space themes. She is a member of the International Association of Astronautical Artists (IAAA), the American Institute of Aeronautics and Astronautics (AIAA), The Lifeboat Foundation, and The International Astronautical Federation's Technical Committee for the Cultural Utilization of Space (ITACCUS). King has published and exhibited in the US, Africa and Europe, most notably as the first Western Artist to exhibit at the Municipal Gallery in Timbuktu, Mali at the turn of the millennium, 2000.